T0335973

THE DYNAMICS
OF FINE POWDERS

Elsevier Handling and Processing of Solids Series

Advisory Editor: R. Clift, Guildford, UK

THE DYNAMICS
OF FINE POWDERS

K. RIETEMA

Europlaan 16, 5691 En Son, The Netherlands

ELSEVIER APPLIED SCIENCE
London and New York

ELSEVIER SCIENCE PUBLISHERS LTD
Crown House, Linton Road, Barking, Essex IG11 8JU, England

Sole Distributor in the USA and Canada
ELSEVIER SCIENCE PUBLISHING CO., INC.
655 Avenue of the Americas, New York, NY 10010, USA

WITH 17 TABLES AND 98 ILLUSTRATIONS

© 1991 ELSEVIER SCIENCE PUBLISHERS LTD

British Library Cataloguing in Publication Data
Rietema, K.
    The dynamics of fine powders.
    1. Powders
    I. Title
    620.43

    ISBN 1-85166-594-3

Library of Congress Cataloging-in-Publication Data
Rietema, K.
    The dynamics of fine powders / K. Rietema.
        p. cm.
    Includes bibliographical references and index.
    ISBN 1-85166-594-3
    1. Powders.   2. Fluidization.   I. Title.
    TP156.P3R54   1991
    620'.43—dc20   91-13476
                                                          CIP

Photoset by Interprint Ltd (Malta).

Preface and Acknowledgements

Aerodynamics and hydrodynamics are well-developed concepts of the relationship between movement and flow of fluid matter, on the one hand, and the forces acting on that matter, on the other hand. Similar concepts on the dynamics of powders, however, are hardly to be found.

Although much research has been done on special aspects of powders, the dynamic behaviour of powders has failed to draw much attention and no systematic study on the dynamics of powders is to be found in the present literature. This is surprising because reliable scaling-up of powder-handling operations seems impossible as long as the dynamics have not been solved. Especially, transfer processes such as heat and mass transfer, as well as mixing and flow of powders, depend strongly on scaling-up and hence on the dynamics.

This book is an account of more than 20 years' research on powders carried out at the Technical University of Eindhoven in the Netherlands. This research started with the discovery that in homogeneous fluidization of fine powders the fluidized particles stay in permanent contact with each other, thus refuting the then general assumption of free floating of fluidized particles. Indeed, theoretical analyses of the stability of gas-fluidized beds which were based on this assumption failed to predict the possibility of stable homogeneous fluidization, thus contradicting experimental evidence of homogeneous fluidization at higher bed expansion.

The logical conclusion was that, owing to the permanent contact between the particles, interparticle forces would exist which had to be introduced in the momentum equations of the theoretical analysis and which were responsible for the stabilization. Later on, the discovery that the viscosity of the fluidization gas has a remarkable effect on the maximum possible bed expansion made it clear that a gas-fluidized system should be considered as a two-phase system of fluidized particles and the fluidization gas. A third important discovery was that interparticle forces are enhanced by adsorption of gas to the surface of the solid particles, thus increasing the maximum possible homogeneous bed expansion at higher gas pressures.

The increased insight into the behaviour of gas-fluidized powders and the effect of the nature of the gas phase on this behaviour suggested also that in other powder-handling operations, such as grinding and mixing, the nature of the gas phase might have a similar strong effect on the behaviour of the powder during operation. This idea was corroborated by an extensive research programme on grinding of powders in ball mills at various gas pressures and with various gases. Also the effect of the gas phase on the rate of mixing in a rotating drum was studied. This was the first systematic research programme on the effect of the gas phase in powder-handling operations known in the literature.

This book is written such that most chapters can be read independently: each chapter has its own introduction and its own notation and list of references.

The research described was carried out in close cooperation with my co-workers, whose names are given here in chronological order: Dr A.A.H. Drinkenburg, Dr R.D. Oltrogge, Dr J.H.B.J. Hoebink, Dr S.M.P. Mutsers, Dr G. Van den Langenberg–Schenk, Dr H.E.A. Van den Akker, Dr G. Van Duyn, Dr A.H.M. Ver-kooyen, Dr E.J.E. Cottaar, and Dr H.W. Piepers. I am grateful for their experimental and theoretical contributions and for the numerous fruitful discussions we had together, which finally resulted in the consistent theory on the dynamics of fine powders presented in this book.

I also wish to acknowledge the contribution of Mr J. Boonstra who carried out many experiments and who did the necessary photographic work. I am very grateful to Mr K.L.M.W. Janssen who took care of and elaborated the nearly 100 graphs and drawings.

Last, but not least, I owe Dr E.D. Kunst a great debt of gratitude. Out of friendship and quite disinterestedly he read the manuscript and rectified the grammatically incorrect passages in the text.

Contents

1

General Introduction

NOTATION

A_p	Surface area of electrodes (m^2)
C	Capacity (F)
C_p	Capacity of a pair of particles (F)
C_s	Capacity of a string of particles (F)
H	Bed height (m)
H_0	Packed bed height (m)
I	Electrical current through powder bed (A)
L	Distance between inserted electrodes (m)
N	Number of parallel strings (m^{-2})
Δt	Time between two collisions (s)
ΔV_0	Potential difference over electrodes (V)
ΔV_s	Potential difference over string of particles (V)
$\bar{d}_p$	Average particle size (m)
i	Electrical current density (A m^{-2})
i_s	Electrical current during a collision (A m^{-2}s^{-1})
l	Distance between electrodes (m)
n	Number of particles in a string (—)
m	Number of collisions per particle per second (s^{-1})
q_n	Electrical charge on a particle (C)
v_{co}	Superficial gas velocity (m s^{-1})

v_{mf}	Superficial gas velocity at minimum fluidization (m s^{-1})
v_{bp}	Superficial gas velocity at which the first bubbles appear (m s^{-1})
α_c	Maximum tilting angle (—)
ε	Porosity of powder bed (—)
ρ_d	Particle density (kg m^{-3})

1.1 POWDERS: WHAT ARE THEY?

Powders appear to be an ill-defined group of substances. The scientific literature on powders does not provide any evidence of what is or should be covered by the term, nor can a clear-cut definition be found. In the large international dictionaries such as the *Encyclopedia Britannica*, the *Encyclopedia Americana*, 'Webster', etc., a powder is stated to be:

(1) matter in a finely divided state: particulate matter;
(2) a preparation in the form of fine particles, especially for medical use;
(3) any of various solid explosives (gun powder).

It is interesting to note that the French *Dictionnaire Encyclopédique Quillet* gives the following definition:

poudre = poussière, petites particules de terre desséché, qui se lèvent au moindre vent.

Only the Dutch *Winkler Prins Encyclopedie* mentions an upper limit of the particle size of the individual particles, viz. 100 μm.
    Probably the general conception of a powder is that of a collection of small discrete solid particles in close contact with each other, the (empty) space between the particles being usually filled with gas so that the bulk density of a powder is always considerably lower than the density of the individual particles. However, this definition also covers a heap of pebbles which no one would call a powder. Apparently a criterion concerning the maximum particle size should be added. If one considers cement,

flour, potato starch, cracking catalyst, sand, and gravel, one will probably agree that the first four materials definitely are powders and the last one certainly is not. Whether one would call sand a powder probably depends on the particle size and on personal views.

When the astronaut Neil Armstrong returned to the Earth from his trip on the surface of the Moon, he stated: 'The surface is fine and powdery. I can kick it up loosely with my toe. It does adhere in fine layers like powdered charcoal to the sole and inside of my boots. I only go in a small fraction of an inch, but I can see the footprints of my boots and the treads in the fine sandy particles.'

These words clearly show that the behaviour of powders depends on the circumstances. In what respects are those on the Moon different from those on the Earth?

(1) The gravitational force on the surface of the Moon is only one-sixth of that on the Earth.
(2) There is no gas on the Moon.

The latter aspect means that any water brought there would evaporate and disappear immediately, hence powders on the Moon will always be perfectly dry so that cohesion between the separate particles due to liquid bridges will be zero. On the other hand the cohesion due to Van der Waals forces remains the same and this means that the ratio of the effective cohesion force to the gravitational force (which in chapter 3 will be called the cohesion number) might not be too different from that of wet sand at the Earth. What Armstrong said about cohesion of the fine sand might point in this direction.

The absence of gas on the Moon, of course, also means that friction forces only result from direct contact between particles. The motion of particles swept up in one way or another is not hindered by viscous friction forces with a surrounding medium.

The subject of this book is confined to the mechanics and dynamics of fine powders in more or less close packings (porosity

<0·7 under terrestrial[†] circumstances) for which the cohesion number is >0·1. In most cases this means that the average particle size is <200 μm.

1.2 POWDER BEHAVIOUR

Powders show a typical and often unpredictable behaviour which is quite different from the behaviour of gases, liquids and solid matter. It has even been suggested that powders constitute a fourth aggregation state. This, however, is nonsense since the thermodynamic properties of powders show a linear relationship with the thermodynamic properties of the component phases: solid and gas.

The behaviour of fine powders under terrestrial circumstances is mainly controlled by two mechanisms:

(1) the particle–particle interaction when the particles are in close contact and which results in friction and cohesion between the particles.
(2) the solid–gas interaction which is of a two-fold nature:
    (a) a hydrodynamic interaction by viscous forces, and
    (b) a physico-chemical interaction via gas adsorption to the solids surface, which at high pressures affects the cohesion forces between the particles.

The importance of solid–gas interaction was mentioned earlier by Bruff and Jenike (1967) and by McDougall (1969), in both cases in connection with solids discharge from hoppers. These two papers failed, however, to draw much notice. Generally, powder scientists did not recognize the effect which ambient gas can have on powder operations. On the other hand investigators of fluidization, in most cases, denied the role of particle–particle interaction during fluidization on the basis of the assumption that while fluidized the solid particles are free floating. A first paper on this subject (Rietema,

---

[†]The term 'terrestrial circumstances' implies that the powder experiences the action of some kind of conservative force (such as the gravitational force) and the presence of a surrounding viscous gas.

1967) was received with much scepticism. Up till now this has led to the remarkable situation that only incidentally has exchange of experience taken place between the two groups of investigators (Rietema, 1984).

One of the most remarkable properties of powders is that they can flow when a certain critical yield stress is exceeded. Many powders even flow quite easily. Before flow sets in from the packed state the powder will expand a little so that the mobility of the separate particles is increased.

It is well known that a paste with the same solids concentration but in which the space between the particles is filled with a liquid, has much poorer flow properties than a powder. This is mainly due to the very low compressibility of a liquid which strongly hinders the necessary dilatation of the particle assembly before flow sets in.

Another striking property of most powders is their capacity to be fluidized by a gas being blown upwards through them at such a rate that they expand. After what has been said above it will be clear that in this fluidized state the higher the expansion, and hence the higher the flow rate of the fluidization gas, the better the powder flows.

A third property of powders is that the particles tend to cohere in a more or less dense swarm which can only relatively slowly be diluted. This swarm coherence is partly due to the cohesion between the particles and is partly the consequence of the fact that dilution of the powder must be accompanied by inward penetration of gas which is hindered by viscous forces and hence creates an outwardly directed pressure gradient. Swarm coherence is also the reason why in fluidization at excessive gas rates a major part of the gas rises rapidly as empty gas envelopes, the so-called bubbles.

1.3 POWDERS IN INDUSTRY

Powders are very frequently met both in daily life and in industry. Because of their use in all domestic housekeeping (sugar, salt, flour, instant coffee, washing powder, etc.) everyone is familiar with powders and their great variation in behaviour. Of much greater economic significance is the production of powders in industry.

The food industry produces many different powders, such as corn starches and potato starches, milk powder and many other spray-dried products. In the pharmaceutical industry many medicines are produced in tablets by compression of powders. The ceramics industry makes all kinds of products via sintering after compression in moulds. The electronics industry applies flourescent and magnetic powders in various widely used products. Electric power plants more and more revert to the burning of powdered coal to produce the necessary thermal energy. The chemical industry, finally, is probably the largest user of powders with its widespread use of catalyst powders, while on the other hand many of the final chemical products, notably plastics and other polymers, are delivered as powders.

1.4 POWDER OPERATIONS

From closer examination of the appearance and the behaviour of powders it can easily be concluded that there is no uniform behaviour of powders. This also follows from the numerous operations carried out with powders:

— storage of powders in hoppers and bins
— transportation of powders from the store to the process apparatus
— grinding or milling of the powder to improve its accessibility for further processing
— mixing of different powders to realize a product of higher quality
— compression of powders in moulds in order to obtain a preformed solid product
— drying of powders at the end of processes in which the separate powder particles are precipitated from a wet suspension

— granulation of powders to obtain larger grains which can be more easily processed
— classification of powders in fractions of different average particle size or density
— fluidization by blowing gas upwards through a powder bed in order to improve the contact between the powder particles and the fluidization gas, e.g. in chemical processes

1.5 ORIGIN AND PRODUCTION OF POWDERS

The origin of powders is mostly of three kinds:

(1) occurrence in living nature such as fine plant seeds, pollen, spores, flour, starch, etc.;
(2) geological: alluvial sands and sediments, dry clays;
(3) industrial production, which is the most frequent source by far.

Industrial production can be further subdivided into four categories:

(1) the small-scale route, especially for medical purposes;
(2) the chemical process route involving chemical precipitation in the liquid phase followed by concentration and finally drying;
(3) the mineral dressing route involving (sea) mining, breaking, crushing, classification or separation, sieving, etc.;
(4) metallurgical processes in which powders generally are an intermediate product which by compression and sintering are transformed into the final product.

The number of different ways of producing powders is surprising. One of the oldest pieces of apparatus used in the production of powders—be it in small quantities only—is the well-known mortar. It was mainly used by alchemists and pharmacists.

Flour is one of the oldest powders produced by grinding of seeds between millstones. The principle of grinding and milling nowadays is further worked out in a whole variety of modern machines.

(a)

(b)

Fig. 1.1. Photographs of 4 powders: (a) spent cracking catalyst, magnification 310 ×; (b) fresh cracking catalyst, magnification 310 ×;

(c)

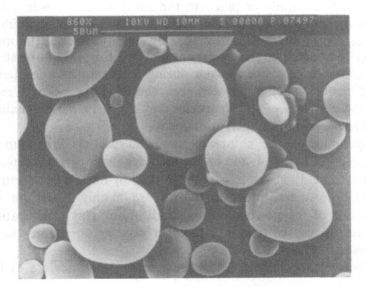

(d)

Fig. 1.1—*contd.* (c) polypropylene, magnification 303 × ; (d) potato starch, *magnification 860 ×.*

Quite different products such as milk powder and cracking catalyst are generally obtained by spray-drying.

Many modern polymers are produced as a suspension of small particles in water, e.g. by suspension polymerization, which after concentration, coagulation and drying yields a powder. Other products are made by precipitation from a solution and subsequent filtration and drying.

Cement owes its constitution to the burning of limestone and clay at high temperatures.

A remarkable powder is potato starch. The individual grains are produced and grow by natural processes in the cells of the potato and have a perfectly smooth surface (see Fig. 1.1). After grating the potato in special machines the starch grains are disclosed and, finally, washed out by some kind of sink-and-float process.

1.6 ABOUT THIS BOOK

This book is mainly confined to the study of fine powders during their treatment in processes such as mixing, grinding, fluidization, etc. In all these processes the powder is in constant motion, generally at some degree of expansion. One can only understand the behaviour of a fine powder during treatment by considering the dynamics of the powder that should be conceived as a two-phase system of solid particles and a gas. In these dynamics the interparticle forces play a major role.

Depending on the nature of these interparticle forces (to be discussed in Chapter 4) and on the fluid dynamics of the two-phase system (to be discussed in Chapters 5 and 6), fine powders can be classified in three categories which are mainly based on the behaviour of the powder during fluidization. This classification, however, is also of relevance in other powder operations (to be discussed in Chapter 11).

According to Geldart (1973) these three categories are as follows:

(1) Category A. Powders that at low gas velocities can be fluidized in a stable expanded state (homogeneous fluidization) but that

at higher gas velocities produce fast rising bubbles (hetero-geneous fluidization).

(2) Category B. Powders that produce bubbles as soon as the gas velocities exceed that at incipient fluidization (= the gas veloc-ity at which the pressure difference over the powder bed has become equal to the weight of the bed divided by the cross-section of the bed). This means that these powders cannot be fluidized homogeneously at the prevalent conditions of gas pressure, temperature and applied fluidization gas.

(3) Category C. Powders that can not really be fluidized and that when the incipient fluidization gas velocity is exceeded develop vertical channels through which most of the gas escapes while the remainder of the powder bed remains stagnant and hardly expands. This phenomenon—called channelling—is generally imputed to excessive high cohesion between the individual particles.

Geldart suggested that the boundaries between these categories can be indicated on basis of particle size and particle density (see Fig. 1.2). In Chapter 6, however, it will be shown that such an approach is inadequate and fails to take the influence of gravity, gas viscosity and cohesion into account.

Original theories on the stability of fluidized powders (Jackson, 1963; Pigford & Baron, 1965) assumed that during fluidization the powder particles are free floating and hence ignored the existence of interparticle forces. These theories predicted that the fluidization would always be of the unstable type (Category B) in contradiction with experimental results. In Chapter 6 it will be shown that when interparticle forces with elastic properties are included in the theoretical analysis the theory does predict homogeneous fluidiz-ation as well as the degree of expansion at which instability sets in.

It has been demonstrated by two different experiments that the particles during fluidization are indeed not free floating but stay in contact with each other.

In the first experiment charcoal particles (as used in coal microphones) were fluidized while two vertical electrodes were immersed in the powder bed. It was found that the bed retained an appreciable electric conductivity at all gas rates (see Fig. 1.3)

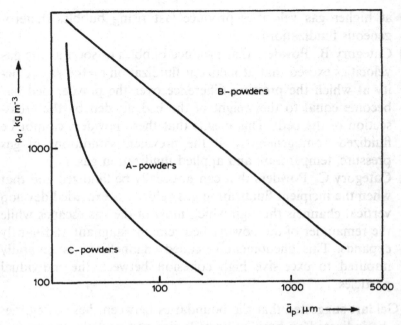

Fig. 1.2. Classification of powders according to Geldart (1973).

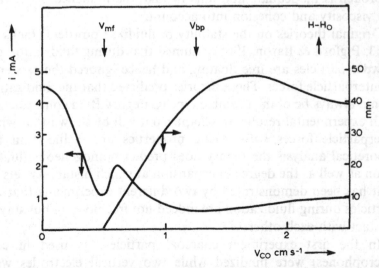

Fig. 1.3. Electrical conductivity and bed expansion of charcoal bed as function of gas velocity.

which is only possible if permanent chains of mutually contacting particles exist in the bed. The electrical conductivity through the bed was indeed highest in the packed bed region and decreased with increasing bed expansion but it was always of considerable magnitude, also in the heterogeneous region (Rietema, 1967). It has been suggested that the conductivity could be explained also by means of frequent collisions between free floating particles while electric charge was transferred from particle to particle at each collision. From a simple theoretical analysis, however, it is shown that the conductivity which follows from such a conception is too small by a factor of 10^6 (see the appendix to this chapter).

In the second experiment a fine powder was fluidized homogeneously in a small bed with a cross-section of $4 \times 10 \, \text{cm}^2$. The bed was then tilted slowly (see Fig. 1.4). It appeared that the bed remained stable while the top surface was tilted as well. If the particles were free floating the top surface would always remain horizontal as in a liquid (Rietema & Mutsers, 1973). This experiment, therefore, demonstrates clearly that the bed must have a mechanical structure with a certain mechanical strength. This can only be caused by linkage between the individual particles due to interparticle forces. When the powder is further expanded homogeneously by increasing the gas rate, some particle contacts must be broken so that rearrangement of the particles results in a looser network with a lower mechanical strength. Accordingly the bed can be tilted less before shearing off sets in. There is a maximum or critical angle α_c over which the bed can be tilted and still retains its structure (no shearing off). As could be expected α_c decreases with increasing expansion and becomes zero when the bed has lost its stability (see Fig. 1.5).

The above makes it clear that from a study of fluidization much can be learned about the properties and the behaviour of powders. It will be shown that these properties and this behaviour are of importance too for powder handling operations in general, especially those operations in which the powder is in constant motion due to increased mobility. The mobility is

(b)

(a)

Fig. 1.4. Tilting fluidized bed: (a) just before and (b) after shearing off.

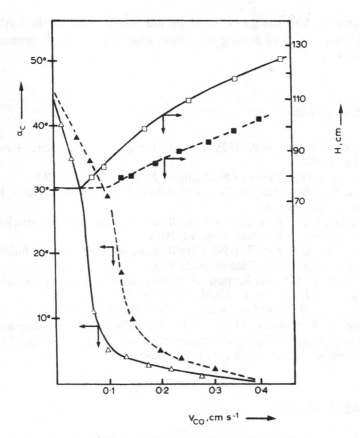

Fig. 1.5. Maximum tilting angle $\alpha_c(\triangle, \blacktriangle)$ and bed expansion $H(\square, \blacksquare)$ versus gas velocity. Open symbols: fluidization with air; solid symbols: fluidization with hydrogen.

increased in most cases by mechanical agitation such as stirring, turning over or reshuffling of the powder. This causes gas to be entrapped so that the powder expands and more or less becomes fluidized. In all cases the fluid dynamic interaction between the solid particles and the interstitial and circumambient gas is of high importance. This means that the powder should be considered as a two-phase system and that properties of the gas phase such as gas viscosity play an important role in the dynamics of powders. Furthermore, a physico-chemical interaction between the particles and the gas caused by gas adsorption to the surface of the particles is of strong influence, especially at high pressures (Chapter 7).

REFERENCES

Bruff, W. & Jenike, A.W. (1967). A silo for ground anthracite. *Powder Techn.*, **1**, 252.

Geldart, D. (1973). Types of fluidization. *Powder Techn.*, **7**, 285.

Jackson, R. (1963). The mechanics of fluidized beds. Part I. *Trans. Inst. Chem. Engrs.*, **41**, 13.

McDougall, I.R. (1969). The ambient fluid influence on solids discharge from hoppers. *Brit. Chem. Eng.*, **14**, 1079.

Pigford, R.L. & Baron, T. (1965). Hydrodynamic stability of a fluidized bed. *Ind. Eng. Chem. Fundamentals*, **4**, 81.

Rietema, K. (1967). Application of mechanical stress theory to fluidization. *Proc. Int. Symp. on Fluidization*, Eindhoven, p. 154.

Rietema, K. (1984. Powders, what are they? *Powder Techn.*, **37**, 5.

Rietema, K. & Mutsers, S.M.P. (1973). The effect of interparticle forces on the expansion of a homogeneous gas fluidized bed. *Proc. Int. Symp. Fluidization and its Applications*, Toulouse, p. 28.

1.A APPENDIX

We wish to derive a rough estimate of electrical conductivity by particle collisions in a gas-fluidized bed of free floating conductive particles.

We consider a suspension of free floating conductive particles in an electrical field between two vertical electrodes. Each pair of two neighbouring particles is conceived as a capacitance. The capacity of a condensor consisting of two parallel electrodes at a distance l is

$$C = \frac{A_p}{4\pi l} \times \frac{10^{-9}}{9} \ \text{F}$$

in which A_p is the surface area of the electrodes.

In the case of the condensor consisting of two neighbouring

particles, the effective area $A_p = \pi d_p^2/4$ and the effective distance $l = \varepsilon d_p/4$. Hence the capacity of a pair of particles becomes

$$C_p = \frac{d_p}{4\varepsilon} \times \frac{10^{-9}}{9} \ \text{F}$$

We now consider a string of n particles in the direction of the lines of force, hence in the direction perpendicular to the electrodes. The capacity C_s of this string, which arises through the cascade of particle capacities in the string, is given by

$$C_s = C_p/n$$

If ΔV_0 is the potential difference between the two electrodes and L is the distance between these electrodes, then the potential difference ΔV_s over the string is

$$\Delta V_s = \Delta V_0 \times \frac{n d_p}{L}$$

By induction from the electrical field there arises on each particle of the string at one side a charge q_n and on the other side a charge $-q_n$, where

$$q_n = \frac{C_p}{n} \Delta V_s$$

When in the string a collision occurs between two neighbouring particles, the number of capacities in the string decreases from n to $(n-1)$ and hence the charge increases from q_n to

$$q_{n-1} = \frac{C_p}{n-1} \Delta V_s$$

Hence an electric charge flows equal to

$$q_{n-1} - q_n = C_p \Delta V_s \left\{ \frac{1}{n-1} - \frac{1}{n} \right\} \approx \frac{C_p \times \Delta V_s}{n^2}$$

When there is a time Δt between two collisions, there arises a small electric current i_s:

$$i_s = \frac{C_p \times \Delta V_s}{n^2 \times \Delta t}$$

Suppose the number of collisions per particle and per second is m. Then the number of collisions per second in a string of n particles is nm, and the time Δt between two collisions in the string is $\Delta t = 1/nm$.

When per unit cross-sectional area there are N parallel particle strings, the electric current density $i = N i_s$ and the electric current through the electrodes will be $I = N i_s A_p$. With $m = v_{co}/d_p$ and $N = (1 - \varepsilon)/d_p^2$ we obtain, after substitution of C_p and ΔV_s,

$$I = \frac{1 - \varepsilon}{4\varepsilon} \left(\frac{u_0}{d_p} \right) \frac{A}{L} \Delta V_0 \frac{10^{-9}}{9} \text{ ampere}$$

With $\varepsilon = 0{\cdot}5$, $A = 10^{-3} \, \text{m}^2$, $v_{co} = 10^{-2} \, \text{m s}^{-1}$ $L = 10^{-1} \, \text{m}$, $d_p = 10^{-4} \, \text{m}$, and $\Delta V_0 = 20$ volt, we obtain $I = 0{\cdot}5 \times 10^{-9}$ ampere. This is smaller by a factor of 10^6 than the measured electrical current through the particle bed.

2

Particle and Powder Characteristics

NOTATION

d	Short axis of ellipsoid (m)
d_p	Particle diameter (m)
d_{pA}	Sauter mean diameter (m)
D	Diameter of container (m)
$f_n(d_p)$	Frequency distribution of particles (m^{-1})
$g(d_p)$	Particle size distribution function on weight basis $(kg\,m^{-1})$
$g'(d_p)$	Normalized weight distribution function (—)
k	Coordination number (—)
K	Permeability (m^2)
l	Long axis of ellipsoid (m)
N	Total number of particles in assembly (—)
S_p	Specific surface area of particle (m^{-1})
W	Total weight of assembly (kg)
r	Radial coordinate (—)
v	Local fluid velocity $(m\,s^{-1})$
$\bar{v}$	Average fluid velocity $(m\,s^{-1})$
δ	Volume fraction of large spheres (—)

ε	External porosity of powder packing (—)
ε_i	Internal porosity of particles (—)
ρ_c	Fluid density (kg m^{-3})
ρ_p	Particle density (kg m^{-3})
ρ_s	Skeletal density (kg m^{-3})
σ_W	Standard deviation (m)
ϕ_A	Relative spread of specific surface area (—)
ϕ_W	Relative spread on weight basis (—)
ψ	Sphericity of particle (—)

2.1 CHARACTERIZATION OF SINGLE PARTICLES

Powders are a dense assembly of discrete particles in a gas. It is obvious, therefore, that a description of powders should start with the characteristics of the individual particles such as diameter, shape, density and surface condition.

Shapes of particles can be very irregular and might vary from small fibrils or rods to small plate-like or spherical particles. This is generally specified by a shape factor, the sphericity ψ which is the ratio of the surface area of a pure spherical particle of the same volume to the real surface area of that particle. This shape factor varies from much smaller than unity for fibrils and rods to nearly unity for most powders. As an example the sphericity of an ellipsoid is $(d/l)^{1/3}$ where d is the short axis of the ellipsoid and l the long axis. It will be clear that possible surface roughness of the particle—which increases the surface area of the particle—should not be taken into account.

The diameter d_p of a particle is generally defined as the diameter of a pure spherical particle with the same volume. This means that for an ellipsoid the particle diameter $d_p = (d^2 l)^{1/3}$.

The above definitions of ψ and d_p only work well as long as the particle shape is not too extreme, which means that the sphericity is not too small, say >0.5. In practice, however, this is not a severe restriction as in most powder operations extreme asperities and protuberances will be ground down or polished by attrition.

In most cases the density of a particle (ρ_p) is simply equal to the density of the solid with the same chemical composition as the particle. There are, however, a number of powders, notably solid catalysts and some polymers, which are produced by agglomeration or coagulation of much smaller particles—often with a diameter $\ll 1\mu m$—and hence look like bunches of grapes through the microscope. Such agglomerates have an internal porosity ε_i which is defined as the ratio of the volume of the internal pores to the external particle volume. The particle density ρ_p is now given by the relation

$$\rho_p = (1 - \varepsilon_i)\rho_s + \varepsilon_i \rho_c$$

in which ρ_s is the density of the solid material, now generally called the skeletal density, while ρ_c is the density of the fluid present in the internal pores of the agglomerate; when the fluid is a gas ρ_c can nearly always be neglected. ε_i may vary from 0·4 to 0·7. A few methods for determining ε_i are given in the appendix to this chapter.

A derived quantity of the particle is its specific surface area S_p which is defined as its real surface area over its real volume. Hence $S_p = 6\psi/d_p$.

2.2 PARTICLE SIZE DISTRIBUTION

The composing particles of an assembly will seldom be of a uniform size. This means that there will be a particle size distribution which can be quantified by a function $f_n(d_p)$ such that $f_n(d_p) \cdot \Delta d_p$ indicates the number of particles of size between d_p and $d_p + \Delta d_p$. When the assembly is large enough f_n may be approximated by a continuous function and in that case it follows that

$$\int_0^\infty f_n(d_p) \cdot \delta(d_p) = N \tag{2.1}$$

where N is the total number of particles in the assembly.

In most cases it is more convenient to define a function $g(d_p)$ such that $g(d_p) \cdot \Delta d_p$ indicates the total weight of particles with size between d_p and $d_p + \Delta d_p$. Now

$$\int_0^\infty g(d_p) \cdot \delta(d_p) = W \tag{2.2}$$

where W is the total weight of the assembly. Of course there is a relation between f_n and g which runs

$$g(d_p) = \frac{\pi}{6} \, \rho_p d_p^3 \cdot f_n(d_p) \tag{2.3}$$

Generally the function g is normalized which means that a new function g' is defined as

$$g'(d_p) = g(d_p)/W$$

so that

$$\int_0^\infty g'(d_p) \cdot \delta(d_p) = 1$$

When g goes to zero only when d_p goes to zero, it follows from eqn (2.3) that it is then possible that f_n goes to infinity and hence N is infinite. When this is the case normalization of f_n is not possible. This is why g or g' is a more workable function than f_n.

An average particle size d_p can be defined on the basis of g' by

$$\bar{d}_{pW} = \int_0^\infty d_p \cdot g'(d_p) \cdot \delta(d_p) \tag{2.4}$$

In processing of a powder the mutual interaction between the particles and the interaction between the particles on the one hand and the interstitial fluid on the other hand are of decisive importance. These interactions can be various:

(1) mechanical friction and cohesion between the particles;
(2) viscous friction between the particles and the fluid when this fluid is in motion relative to the particles;
(3) buoyancy operating on the particles;
(4) adsorption of chemical components in the fluid at the surface of the particles;

(5) chemical reaction between fluid components and the solid.

All of these interactions (except the buoyancy) operate at the surface of the particles. This is the reason why a second average particle size based on the specific surface area of the powder S_A is defined, the so-called Sauter mean diameter d_{pA}:

$$\bar{d}_{pA} = \frac{6\psi}{\bar{S}_A}$$

$$\bar{S}_A = \int_0^\infty \left(\frac{6\psi}{d_p}\right) \cdot g'(d_p) \cdot \delta(d_p) \tag{2.5}$$

It is this average diameter (indicated in the following by $\bar{d}_p$) that will mostly be used in this book.

Besides these averages other important parameters of a distribution function are the variance and two quantities derived from it, the spread or standard deviation and the relative spread. Of the distribution function g' with average parameter $\bar{d}_{pW}$ the variance σ_W^2 is given by the equation

$$\sigma_W^2 = \int_0^\infty (d_p - \bar{d}_{pW})^2 \cdot g'(d_p) \cdot \delta(d_p) \tag{2.6}$$

With $\overline{d_p^2} = \int_0^\infty d_p^2 \cdot g'(d_p) \cdot \delta(d_p)$ it can easily be shown that

$$\sigma_W^2 = \overline{d_p^2} - (\bar{d}_{pW})^2 \tag{2.7}$$

In these equations σ_W is the standard deviation. The relative spread ϕ_W is given by

$$\phi_W = \frac{\sigma_W}{\bar{d}_{pW}} \tag{2.8}$$

When the Sauter mean diameter of the distribution is used, also the variance and the standard deviation should be based on the specific surface area. Hence

$$\sigma_A^2 = \int_0^\infty \left(\frac{1}{d_p} - \frac{1}{\bar{d}_p}\right)^2 \cdot g'(d_p) \cdot \delta(d_p) \tag{2.9}$$

The relative spread ϕ_A now becomes

$$\phi_A = \frac{\sigma_A}{1/\overline{d_p}} = \overline{d_p} \sqrt{\overline{\left(\frac{1}{d_p^2}\right)} - \frac{1}{\overline{d_p}^2}} \qquad (2.10)$$

in which

$$\overline{\left(\frac{1}{d_p^2}\right)} = \int_0^\infty \frac{1}{d_p^2} \cdot g'(d_p) \cdot \delta(d_p) \qquad (2.11)$$

Care should be taken not to make the error of putting $\overline{(1/d_p^2)}$ equal to $1/\overline{d_p}^2$.

Methods of determining the particle size distribution are:

(a) dry sieving
(b) sedimentation
(c) the Coulter-counter principle
(d) light scattering
(e) microscopy.

Methods (a) and (b) determine the distribution function g and methods (c), (d) and (e) determine the distribution function f_n. It must be realized that the results of these methods are not always comparable except when the particles are purely spherical.

Method (a), sieving, is generally performed in a stack of sieves of decreasing mesh number (increasing aperture) from the bottom upwards. The whole stack is vibrated mechanically so that the powder is classified in fractions of particles between two sizes. After separation the fractions are collected and weighed, thus giving information on the distribution function g. Sieving is generally not adequate to give information below a particle size of 40 μm because of cohesion between the particles or adhesion of the particles to the fine wires of the mesh. Sieving is the only method that is operated with the dry powder. All other methods are operated with the powder suspended in a suitable liquid that wets the particles completely.

Method (b), analysis by sedimentation, is based on the dependence of the settling rate on the size of the particles. In order to prevent mutual hindrance of the individual particles, the analysis has to be done at sufficient low concentrations (<1% by volume). Three methods are used to carry out the sedimentation analysis:

(1) continuous weighing of a small scale near the bottom of the sedimentation column whose weight increases in time due to the settling of the particles (sedimentation balance);
(2) taking samples by pipetting at different times at a specific constant height and measuring the total weight of the collected particles (Andreasen method);
(3) measuring the decrease of the hydrostatic pressure with time at a small distance from the bottom of the sedimentation column, which decrease is due to the decreasing average concentration of the suspension (Wiegner method).

Method (c), the Coulter-counter principle, determines in succession the size of each particle in a very diluted suspension of the powder in an electrolyte while the suspension is sucked through a small orifice on either side of which an electrode is immersed. The change of electrical resistance between the electrodes is a measure of the volume of the particle. Thus the number of particles which fall in a certain size class can be counted. Of course the concentration of the particles in the electrolyte has to be so low that the chance that more than one particle at a time passes through the orifice can be neglected. This means that only a very small sample of the powder can be brought in. Accordingly special attention must be paid to ensure that the sample is representative of the powder as a whole.

Method (d), the light scattering method, is founded on the principle that a beam of monochromatic light falling on a particle is scattered at a specific angle, which is a measure of the particle size. From the intensity of the scattered light as a function of the scattering angle the particle size distribution of a suspension can be derived. Applying specially designed photometers which use a laser as the light source permits a very effective and rapid analysis of the powder.

In Method (e), a small sample of powder after immersion in a wetting liquid is spread on a microscope slide. On an enlarged photograph the separate particles can be counted and their size determined by means of scanning with a computer. Microscopy is the only method that also gives information on the shape of the particles, although it must be realized that only a two-dimensional representation of the particles is given. For more details readers are referred to the literature.

2.3 THE POWDER PACKING

If the particles of an assembly are not being blown upwards, nor blown away, nor are falling freely, they generally compose what is called a packing in which the individual particles are piled up, resting on each other in the gravitational field. As the particles do not occupy all of the space in the packing there remains an 'empty' space. The volume of this empty space divided by the total volume of the packing is the so-called external porosity ε (also often called the voidage). Together with the internal porosity ε_i (if any) of the particles, the total porosity of the assembly becomes $\varepsilon_t = \varepsilon + (1 - \varepsilon)\varepsilon_i$.

The external specific surface area of the packing is reduced by a factor $(1 - \varepsilon)$ so that now $\bar{S} = \bar{S}_A(1 - \varepsilon) = 6(1 - \varepsilon)\psi/\bar{d}_p$.

Directly related to the external porosity is the permeability K of the packing, i.e. the capacity of the packing to let a fluid pass under the influence of a pressure gradient (see Chapter 5). The permeability is strongly dependent on the porosity and is further determined by the specific surface area of the packing: $K \sim \varepsilon^3/\bar{s}^2$.

When the particles are perfect spheres of identical size one can imagine a so-called closest packing or rhombohedral packing with a porosity $\varepsilon = 0.2595$. In contrast the cubic packing of uniform spheres has a porosity $\varepsilon = 0.4764$. These porosities, however, have only theoretical value as in practice there are strong deviations due to the following four causes:

(a) the wall effect
(b) the effect of random settling and friction
(c) the influence of cohesion between the particles
(d) the influence of particle size distribution.

In most cases the powder is present in a container confined by walls. Since particles cannot lay partly inside and partly outside the container, some space is regularly left open near the wall that would otherwise be occupied by a particle (see Fig. 2.1). Owing to

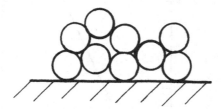

Fig. 2.1. Increased porosity near the wall.

this effect the regular packing near the wall and at some distance from the wall is disturbed. A consequence of this disturbance is that the permeability near the wall is increased. Schwartz and Smith (1953) determined this effect experimentally. Their results are shown in Figs 2.2 and 2.3. In Fig. 2.2 the local fluid velocity v—as ratio of the average fluid velocity $\bar{v}$—is shown as a function of the distance r from the centre of the container for two ratios of the particle size d_p to the container diameter D. Figure 2.3 shows the contribution of this increased permeability as the percentage of the total throughput in a cylindrical container as a function of

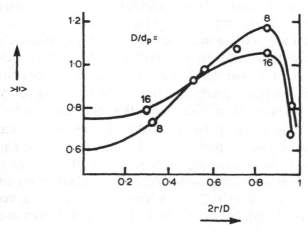

Fig. 2.2. Local fluid velocity near the wall of the container.

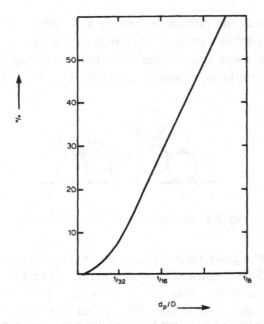

Fig. 2.3. Contribution of increased permeability near the wall as percentage of total throughput in a cylindrical container.

d_p/D. From this figure it is clear that in the case of fine powders this effect can be neglected.

The rhombohedral and cubic packings (and other regular packings) can be attained only by carefully piling up the particles by hand while in practice the packing is the result of a random process, usually settling under influence of a gravitational or other conservative force field. In this settling process the particles move downwards in a disorderly fashion owing to mutual hindrance. When the particles reach the top layer of the already settled material mechanical friction with this top layer may prevent further settling of the particles so that the disorder is maintained in the build-up of the packing, resulting in an increased porosity.

The third effect, cause (c), is the result of attractive or repulsive forces between the particles. Attractive forces (cohesion) are mostly due to Van der Waals forces, which work only at a very short range ($<10^{-3}$ μm), but on this short range these forces can be very strong. When the particles are not completely dry the attraction may be the result of capillary forces. Repulsive forces may originate from electrostatic charging of the particles which can occur by friction or collision of the particles with stationary walls of a dissimilar material (triboelectrification). These forces, however, are much smaller than the Van der Waals forces.

Especially for fine powders ($d_p < 100\,\mu m$) the Van der Waals forces can be very dominant and much higher than the weight of the individual particles. In Chapter 4 these forces will be discussed in more detail.

When an attractive force exists between the particles a settling particle will become stagnant as soon as it makes contact with an already settled particle in the top layer of the packing. Thus the settling particle will be prevented from rolling into a neighbouring open space. As a result the final packing will contain many open cavities, which contribute to an increased porosity.

The reverse phenomenon might occur when there is a repulsive force between the particles: the settling particles are directed towards open cavities by these repulsive forces. This will result in a lower porosity of the final packing.

Similar phenomena can be observed during settling of PVC particles in water. The PVC particles are poorly wetted by water so that aggregates of particles are formed during settling, which results in a high porosity of the final packing. By addition of a suitable surface active agent the wetting can be much improved. The agent is adsorbed to the surface of the particles which causes a very thin electrical double layer to be formed around the particles. Interaction of these double layers—when the particles approach each other—now causes the particles to repel each other (Verwey & Overbeek, 1948) and the settled layer attains a porosity that is even lower than the porosity attained if the same particles settle in air.

With regard to causes (b) and (c), vibration of the container at a moderate frequency or tapping the container can break the bonds between the individual particles so that the structure of the packing is continuously disturbed and gravitational force can cause the packing to become more dense.

Particle size distribution, cause (d), in most cases will lead to an increased packing density (lower porosity) especially when there is a strong relative spread in particle size. Many authors (Fraser, 1935; Eastwood *et al.*, 1969; Rodriguez *et al.*, 1986; Stovall *et al.*, 1986) have investigated this phenomenon, mostly by studying binary mixtures of large and small spheres.

If the quantity of small spheres is so high that the large spheres 'swim' in the structure of small spheres and do not touch each other (see Fig. 2.4) the overall porosity is mainly determined by the

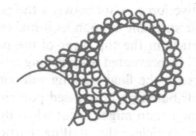

Fig. 2.4. Two-component mixture of particles at low volume fraction of large spheres.

small spheres. When δ is the volume fraction of large spheres and ε_1 is the porosity of the suspension of small spheres, the overall porosity $\varepsilon = \varepsilon_1(1 - \delta)$. As the number of large spheres increases the porosity will initially decrease. If the ratio of the large sphere diameter d_1 to the small sphere diameter d_2 is large enough $(d_1/d_2 > 5)$ then at a specific volume fraction of large spheres these spheres will touch each other and form a packing structure of their own with the small spheres filling the pores between the large spheres (see Fig. 2.5). In this special case the overall porosity will be at a minimum.

When the volume fraction of large spheres is further increased the small spheres will not be enough to fill up all of the pores between the large spheres so that the porosity will increase again (see Fig. 2.6). Figure 2.7 is a plot of the overall porosity versus the volume fraction of large spheres (these figures are taken from Fraser, 1935).

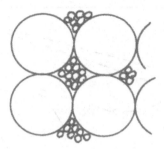

Fig. 2.5. Two-component mixture of particles. Large spheres touching. Large pores completely filled by small spheres.

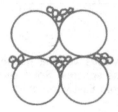

Fig. 2.6. Two-component mixture of particles. Large spheres touching. Large pores partly filled by small particles.

When cohesion and friction between the particles can be neglected the porosity should in principle be the same at zero and unity volume fraction of large spheres. That this is not so in Fig. 2.7 is probably caused by the wall effect mentioned under (a).

In multi-component mixtures the effect of a particle size distribution is even more complicated but here too the overall effect will be of a decreased porosity. On the other hand, in such systems stable structures at increased porosity are also possible when cohesion and friction are relatively strong and cannot be neglected. The principle of this phenomenon is illustrated with the aid of a two-dimensional presentation of a 'powder' consisting of an eight-component mixture of flat coins (see Fig. 2.8(a) and (b)). In this figure two structures of the same assembly are shown: in Fig. 2.8(a) a close packing with a 'porosity'

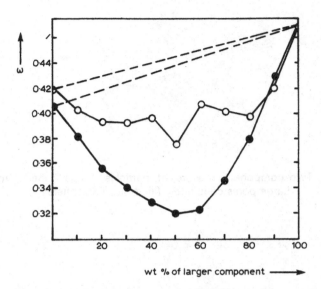

Fig. 2.7. Overall porosity versus volume fraction of large spheres. Open symbols: 5 and 16 mm glass spheres. Solid symbols: 3 and 16 mm glass spheres. Results taken from Eastwood *et al.* (1969) with permission.

$\varepsilon = 0.162$ and in Fig. 2.8(b) a loose packing with a 'porosity' $\varepsilon = 0.366$. This latter packing might arise when high cohesion and friction prevent particles from shearing off to sideward positions so that piling up in the vertical direction becomes possible with relatively little sideward support, leaving more open cavities. For comparison it may be noted that the porosity of a closest packing of uniform cylinders amounts to 0.0931 and that of a square packing to 0.2146.

A final remark must be made on the fractional open area in a cross-section through the packing. Consider a horizontal slice of the packing with a cross-sectional area A and extending from height h to $h + \Delta h$. When the packing is purely random the average porosity should be everywhere the same. Hence the total open space in the slice is $\varepsilon A \Delta h$. The total open area in the cross-section at height h is $\varepsilon_a A$ and is the same at the cross-section $h + \Delta h$ and at any cross-section in between. This means that in any slice with infinitesimal thickness dh the

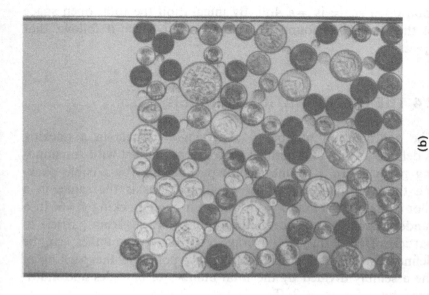

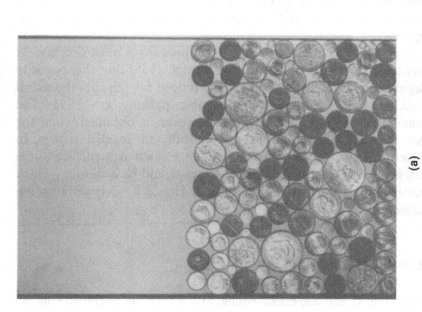

(a)

(b)

Fig. 2.8. 'Two-dimensional powder' (a) at close packing, $\epsilon = 0.162$; (b) at loose packing, $\epsilon = 0.366$.

total open space is $= \varepsilon_a A dh$. By integration the total open space of the slice Δh is found to be $\varepsilon_a A \Delta h$ and hence it follows that $\varepsilon_a = \varepsilon$.

2.4 THE COORDINATION NUMBER

The coordination number k of a single particle in a packing is defined as the total number of points of contact with surrounding particles. In a rhombohedral packing and in a cubic packing the coordination number of all particles is the same: in a rhombohedral packing $k = 12$, and in a cubic packing $k = 6$. In a random packing the coordination number varies from particle to particle. Hence only an average coordination number can be defined as the total number of contact points in the packing of the assembly divided by the total number of particles and multiplied by a factor of 2. This factor of 2 comes in because every contact contributes two contact points, one on each contacting particle.

It will be clear that with increasing porosity k must decrease. It has been suggested by Smith (see Rumpf, 1958) that this porosity dependence is described by $k\varepsilon = 3\cdot 1$. Indeed for the rhombohedral packing $k\varepsilon = 3\cdot 114$ and for the cubic packing $k\varepsilon = 2\cdot 858$. The maximum porosity of a packing, however, is obtained when the particles are all stacked purely vertically in parallel strings, in which case $k = 2$. In order to stabilize such a structure some sideward support between these strings would be necessary, which means that k must be larger than 2. Hence $k = 2$ is an absolute minimum.

REFERENCES

Eastwood, J., Matzen, E.J.P., Young, M.J. & Epstein, N. (1969). Random loose porosity of packed beds. *Brit. Chem. Eng.*, **14**, 1542.
Fraser, H.J. (1935). Porosity and permeability of clastic sediments. *J. Geol.*, **43**, 785.

Rodriguez, J., Allibert, C.H. & Chaix, J.M. (1986). A computer method for random packing of spheres of unequal size. *Powder Techn.*, **47**, 25.

Rumpf, H. (1958). Grundlagen und Methoden des Granulierens. *Chem.-Ing. Techn.*, **30**, 144.

Schwartz, C.E. & Smith, J.M. (1953). Flow distribution in packed beds. *Ind. Eng. Chem.*, **45**, 1209.

Stovall, T., de Larrard, F. & Buil, M. (1986). Linear packing density model of grain mixtures. *Powder Techn.*, **48**, 1.

Verwey, E.J. & Overbeek, J.Th.G. (1948). *Theory of the Stability of Lyophobic Colloids*. Elsevier, Amsterdam.

2.A APPENDIX. MEASUREMENT OF THE INTERNAL POROSITY OF POWDERS

2.A.1 The BET Method

A powder sample of known weight is brought in a glass container and heated in vacuum to release all gas fractions which might be adsorbed on the powder. Next the container and a similar but empty one are submerged in a bath filled with liquid nitrogen at a temperature of $-195°C$. The two containers are connected with each other by means of a differential manometer. The two containers are evacuated and brought in contact with a known volume of nitrogen. Because of adsorption of nitrogen to the surface of the powder sample, a pressure difference will arise between the two containers which is measured with the differential manometer. From this difference in pressure the amount of nitrogen adsorbed can be calculated. In normal operation this method is used to derive the total surface area of the powder. If the sample is not too small the amount of nitrogen adsorbed will be proportional to the nitrogen pressure above the sample. When, however, the nitrogen pressure is further increased more gas will be adsorbed also on those parts of the total surface that has not yet been covered with adsorbed nitrogen. At still further increase of the nitrogen pressure more monolayers of nitrogen are adsorbed, and finally the internal pores of the powder will be completely filled with adsorbed nitrogen, so that the curve of adsorbed gas versus the gas pressure will flatten off when this saturation value is reached. Hence, from this saturation value the volume of internal pores and the internal porosity can be derived.

2.A.2 The Centrifugal Method

Two glass containers with a porous bottom are filled with equal samples of the powder. The powders are saturated with a wetting liquid and then brought opposite each other in a laboratory centrifuge. The density of the liquid must be lower than the skeletal density of the powder. The centrifuge is started and due to the centrifugal action part of the liquid is filtered off. The amount filtered off depends on the centrifugal force and on the pore size. The larger pores will be emptied first. At increasing rate of revolution of the centrifuge the smaller pores also will be emptied. The revolution rate is increased in steps and after each step the weight of the containers is measured. When this weight is plotted against the revolution rate a curve arises with initially a steep slope until all external pores are emptied. Because the internal pores have a pore size which is a few orders of magnitude smaller than the size of the external pores, the slope of the above curve will strongly decrease with further increase of the revolution rate. Finally, all pores, both external and internal, will be emptied. Of course, the weight of the containers must be known so that the weight m_0 of the dry powder samples can be calculated. By extrapolation of the second part of the curve towards zero revolutions the weight m_1 of liquid in the internal pores can be derived. When the skeletal density ρ_s of the powder and the density ρ_l of the liquid is known, the internal porosity of the powder can be calculated from

$$\varepsilon_i = \frac{\rho_s m_1}{\rho_s m_1 + \rho_l m_0}$$

2.A.3 The Permeability Method

This method is based on the assumption that the pores of the internal porosity are so small compared to the external pores that their contribution to the permeability of the powder bed can be neglected.

In a small fluidized bed, say 4 cm in diameter, a powder sample is fluidized homogeneously. The bed height should be about 10 cm. The fluidization is stopped and the bed is allowed to settle. The packed bed height H_1 is measured. At a gas velocity v_1 below the incipient fluidization velocity the pressure drop Δp_1 over the bed is measured as well. The permeability K_1 is now expressed as $K_1 = H_1 v_1 / \Delta p_1$.

By vibration of the bed the bed height is decreased as well as the external bed porosity. Again the bed height H_2 and the pressure drop Δp_2 are measured at a gas velocity v_2. The permeability $K_2 = H_2 v_2 / \Delta p_2$.

From the Kozeny relation it follows that at constant particle size and constant gas viscosity the permeability K is proportional to $\varepsilon^3 / (1-\varepsilon)^2$ where ε is the external porosity. From mass balance it follows that

$$(1-\varepsilon_1)H_1 = (1-\varepsilon_2)H_2$$

It can now be derived that

$$K_1/K_2 = (\varepsilon_1/\varepsilon_2)^3 \cdot (H_2/H_1)^2$$

Hence the ratio $\varepsilon_1/\varepsilon_2$ can be calculated.

When we put $\lambda = H_1/H_2$ and $\phi = \varepsilon_1/\varepsilon_2$, then it can be derived that

$$\varepsilon_1 = \frac{1-\lambda}{\dfrac{1}{\phi} - \lambda}$$

and

$$\varepsilon_2 = \frac{1-\lambda}{1-\phi\lambda}$$

With W as the total weight of the powder sample, A the cross-sectional area of the bed and ρ_s the skeletal density of the powder, it follows for the internal porosity ε_i that

$$1 - \varepsilon_i = \frac{W/AH_1\rho_s}{1 - \varepsilon_1} = \frac{W/AH_2\rho_s}{1 - \varepsilon_2}$$

3

Mechanics of Powders at Rest

NOTATION

A	$\bar{\rho}g + \mathrm{d}p/\mathrm{d}y$ (see eqn (3.31)) (N m^{-3})
D_c	Maximum diameter of free arch (m)
H	Bed height (m)
I_1, I_2, I_3	Invariants (see eqn (3.16)) (N m^{-2})
l, m, n	Direction cosines (—)
p	Gas pressure in fluidization (N m^{-2})
P	Stress on a plane through O (N m^{-2})
P_d	Isotropic pressure in powder mass (N m^{-2})
q	Stress with components q_x, q_y and q_z (N m^{-2})
r	Radius of Mohr circles
r_a	Radius of critical Mohr circle (N m^{-2})
R	Bed radius (m)
v_{co}	Superficial gas flow rate (m s^{-1})
ε	Porosity of powder bed (—)
β	Slope of heap of powder (—)
θ	Angle between principal axis and x-direction (—)
$\bar{\rho}$	Average bulk density of powder (kg m^{-3})
σ_c	Tension cutoff (N m^{-2})

σ	Powder stress tensor ($N\,m^{-2}$)
$\sigma_1, \sigma_2, \sigma_3$	Principal stresses ($N\,m^{-2}$)
τ	Shear stress ($N\,m^{-2}$)
τ_0	Cohesion constant ($N\,m^{-2}$)
ϕ	Angle of internal friction (—)

3.1 THE STRESS TENSOR

As mentioned in Section 1.6, a powder at rest constitutes a mechanical structure. This means that it can withstand certain stresses applied from the outside without permanent distortion, so that it returns to its original configuration when the stresses are relieved. Thus the powder structure is elastic within certain limits, and inside the powder structure stresses are generated that compensate for the outward stresses. When the elasticity limits are exceeded a plastic deformation results from which the powder will not return to its original configuration. At any point in the powder the stress situation is described by the stress tensor σ. From point to point this stress tensor may vary but only continuously.

Let us consider the stress tensor at a point O with coordinates x, y and z in a rectangular coordinate system. Since stresses always act on a surface we further consider a small cube $(\Delta x, \Delta y, \Delta z)$ at the point O with edges r_0. We choose r_0 so small that across a face of the cube the stress may be assumed constant (see Fig. 3.1). On the face $(\Delta y, \Delta z)$ there acts a force $q = r_0^2 . \sigma$ where σ can be decomposed into a normal stress σ_{xx} and a shear

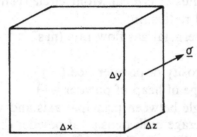

Fig. 3.1. Cube of powder with stress σ on Δy–Δz plane.

stress τ which can be further decomposed into the shear stresses σ_{xy} and σ_{xz}. Similar stresses act on the other faces of the cube. Hence the stress tensor σ has nine components and is accordingly denoted by

$$\begin{pmatrix} \sigma_{xx} & \sigma_{xy} & \sigma_{xz} \\ \sigma_{yx} & \sigma_{yy} & \sigma_{yz} \\ \sigma_{zx} & \sigma_{zy} & \sigma_{zz} \end{pmatrix}$$

On a thin slice dz perpendicular to the z-axis acts a couple due to the shear stresses σ_{xy} and σ_{yx}. As the powder is at rest and will not rotate around the z-axis, this couple must vanish and hence $\sigma_{xy} = \sigma_{yx}$. In the same way it can be argued that $\sigma_{xz} = \sigma_{zx}$ and $\sigma_{yz} = \sigma_{zy}$. A tensor with these properties is called *symmetrical*. On rotation of the coordinate system the stresses, of course, will vary. For a symmetrical stress tensor a certain coordinate system (x', y', z') can thus be found for which the shear stresses $\sigma_{x'z'} = \sigma_{z'x'} = 0$, $\sigma_{x'y'} = \sigma_{y'x'} = 0$ and $\sigma_{y'z'} = \sigma_{z'y'} = 0$ so that only three normal stresses $\sigma_{x'x'}, \sigma_{y'y'}$ and $\sigma_{z'z'}$ remain. These are called the *principal stresses* σ_1, σ_2 and σ_3 respectively, while the axes x', y' and z' are called the *principal axes*. Furthermore it can be shown that $\sigma_1 + \sigma_2 + \sigma_3 = \sigma_{xx} + \sigma_{yy} + \sigma_{zz} = $ invariant. The proof of this will be given in Section 3.3.

3.2 TWO-DIMENSIONAL SYSTEM

We shall first discuss a two-dimentional system in which the stresses depend only on the x- and y-coordinates. As an example one may think of a long dyke in the z-direction with a constant cross-section. In such a system the stresses will be independent of z so that the stress tensor σ is now reduced to

$$\begin{pmatrix} \sigma_{xx} & \sigma_{xy} \\ \sigma_{yx} & \sigma_{yy} \end{pmatrix}$$

with again $\sigma_{xy} = \sigma_{yx}$.

We consider a triangle ABC in a cross-section which again is so small that the stresses may be assumed constant. In fact each side of this triangle represents a plane through this side and extending in the z-direction (see Fig. 3.2). The triangle is chosen such that on the sides AC and BC the principal stresses σ_1 and σ_2

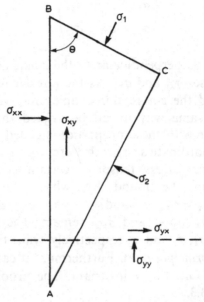

Fig. 3.2. Two-dimensional triangle for analysis of force equilibrium.

($\sigma_1 > \sigma_2$) operate, while on the side AB in the y-direction a normal stress σ_{xx} and a shear stress σ_{xy} operate. On a plane perpendicular to AB there operate the normal stress σ_{yy} and the shear stress $\sigma_{yx} = \sigma_{xy}$. The x-direction makes an angle θ with the first principal axis. Drawing up a force balance on the triangle in the x-direction gives

$$AB(\sigma_{xx}) = BC(\sigma_1 \cos \theta) + AC(\sigma_2 \sin \theta)$$

Since $BC = AB \cos \theta$ and $AC = AB \sin \theta$ it is found that

$$\sigma_{xx} = \sigma_1 \cos^2\theta + \sigma_2 \sin^2\theta$$
$$= \tfrac{1}{2}(\sigma_1 + \sigma_2) + \tfrac{1}{2}(\sigma_1 - \sigma_2) \cos 2\theta \qquad (3.1)$$

From a force balance in the y-direction it is found that

$$\sigma_{xy} = \tfrac{1}{2}(\sigma_1 - \sigma_2) \sin 2\theta \qquad (3.2)$$

Similarly from force balances on a triangle AB′C′ it is found that

$$\sigma_{yy} = \tfrac{1}{2}(\sigma_1 + \sigma_2) - \tfrac{1}{2}(\sigma_1 - \sigma_2) \cos 2\theta \qquad (3.3)$$

$$\sigma_{xy} = \tfrac{1}{2}(\sigma_1 - \sigma_2) \sin 2\theta = \sigma_{yx} \qquad (3.4)$$

It also follows from these relations that

$$\sigma_{xx} \cdot \sigma_{yy} = \sigma_1 \cdot \sigma_2 + \sigma_{xy}^2 \qquad (3.5)$$

$$\sigma_{xx} + \sigma_{yy} = \sigma_1 + \sigma_2 = 2P_d \qquad (3.6)$$

The constant P_d is the so-called isotropic or hydrostatic pressure.

The above relations (3.1)–(3.4) can be plotted geometrically by means of a Mohr-circle, so-called after Mohr who first presented this analysis (see Fig. 3.3). The points on this circle represent the possible stress combinations $(\sigma_{xx}, \sigma_{yy}, \sigma_{xy})$ on different planes through the point A of the powder mass. To each point in the powder mass a specific Mohr-circle corresponds.

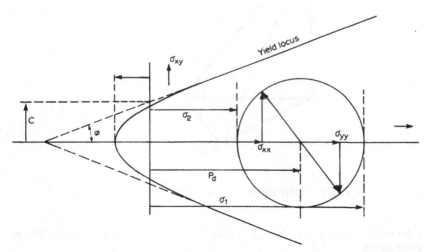

Fig. 3.3. Two-dimensional Mohr circle.

3.3 STRESS ANALYSIS OF THREE-DIMENSIONAL SYSTEMS†

We return to the three-dimensional stress tensor discussed in Section 3.1 and investigate the forces on an arbitrary plane through the point O. The normal on this plane has the direction cosines l, m and n with respect to the axes Ox, Oy and Oz. The direction cosines satisfy the equation $l^2 + m^2 + n^2 = 1$.

We consider the force equilibrium on the small tetrahedron OABC of which the face ABC is normal to the direction (l, m, n) (see Fig. 3.4). Again we consider this tetrahedron to be so small that the stress tensor σ may be considered constant throughout its volume.

If the surface area of ABC is denoted by S, the surface areas of OAB, OBC and OCA are Sn, Sl and Sm respectively. On the

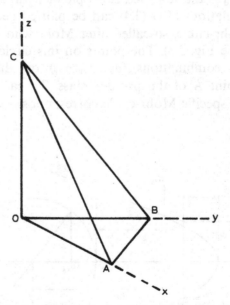

Fig. 3.4. Tetrahedron for analysis of force equilibrium.

†The main lines of the analysis presented in this section follow the derivation as given by Jaeger (1969).

face ABC there operates a stress q with components q_x, q_y and q_z. Force equilibrium on the tetrahedron in the x-direction then requires that

$$\left.\begin{aligned}
Sq_x &= Sl\sigma_{xx} + Sm\sigma_{xy} + Sn\sigma_{xz} \\
\text{or} \quad q_x &= l\sigma_{xx} + m\sigma_{xy} + n\sigma_{xz} \\
\text{and similarly} \quad q_y &= m\sigma_{yy} + l\sigma_{xy} + n\sigma_{yz} \\
q_z &= n\sigma_{zz} + l\sigma_{xz} + m\sigma_{yz}
\end{aligned}\right\} \tag{3.7}$$

When σ is the normal stress on the face ABC then

$$\begin{aligned}
\sigma &= lq_x + mq_y + nq_z \\
&= l^2\sigma_{xx} + m^2\sigma_{yy} + n^2\sigma_{zz} + 2mn\sigma_{yz} + 2ln\sigma_{xz} + 2lm\sigma_{xy}
\end{aligned} \tag{3.8}$$

When we vary the direction (l, m, n) continuously then the normal stress σ will vary too, but there must be a certain direction (l_1, m_1, n_1) for which σ will have a maximum value, and a direction (l_2, m_2, n_2) for which σ is a minimum.

For these two extremes the following conditions must hold:

$$\left(\frac{\partial\sigma}{\partial l}\right)_m = 0 \quad \text{and} \quad \left(\frac{\partial\sigma}{\partial m}\right)_l = 0 \tag{3.9}$$

When the direction cosines l and m are varied independently, it follows from $l^2 + m^2 + n^2 = 1$ that there is no longer a free choice of n. When we differentiate this equation with respect to l and m respectively, it is found that

$$l + n\left(\frac{\partial n}{\partial l}\right)_m = 0 \quad \text{and} \quad m + n\left(\frac{\partial n}{\partial m}\right)_l = 0 \tag{3.10}$$

From eqns (3.8) and (3.9), and simplifying with the help of eqns (3.7), it follows that in the extreme situations

$$q_x + q_z\left(\frac{\partial n}{\partial l}\right)_m = 0 \quad \text{and} \quad q_y + q_z\left(\frac{\partial n}{\partial m}\right)_l = 0 \tag{3.11}$$

When next $(\partial n/\partial l)_m$ and $(\partial n/\partial m)_l$ are eliminated from eqns (3.11), using eqns (3.10) it is finally found that for the extreme values of σ

$$q_x/l = q_y/m = q_z/n \qquad (3.12)$$

This means that the stress q now has the same direction as the normal on the face ABC and hence that q is a normal stress and must be equal to σ. It follows also that for these specific directions there is no shear stress acting on the face ABC. Again these directions are called the principal axes and the corresponding stresses the principal stresses. From eqns (3.12) it also follows that $q_x = l\sigma$, $q_y = m\sigma$ and $q_z = n\sigma$. Combination with eqns (3.7) gives

$$\left. \begin{array}{l} l(\sigma_{xx} - \sigma) + m\sigma_{xy} + n\sigma_{xz} = 0 \\ l\sigma_{xy} + m(\sigma_{yy} - \sigma) + n\sigma_{yz} = 0 \\ l\sigma_{xz} + m\sigma_{yz} + n(\sigma_{zz} - \sigma) = 0 \end{array} \right\} \qquad (3.13)$$

These three equations in l, m and n have three solutions for which the determinant

$$\begin{vmatrix} (\sigma_{xx} - \sigma) & \sigma_{xy} & \sigma_{xz} \\ \sigma_{xy} & (\sigma_{yy} - \sigma) & \sigma_{yz} \\ \sigma_{xz} & \sigma_{yz} & (\sigma_{zz} - \sigma) \end{vmatrix} = 0 \qquad (3.14)$$

The three solutions are σ_1, σ_2 and σ_3 each with a specific direction (l, m, n). These three directions are mutually perpendicular. This can be shown by multiplying the three equations (3.13) now containing σ_1, l_1, m_1 and n_1 by l_2, m_2 and n_2 respectively and subsequently adding them. When next the indices 1 and 2 are interchanged and the second sum is subtracted from the first sum, it is found after some rearranging that

$$(\sigma_1 - \sigma_2)(l_1 l_2 + m_1 m_2 + n_1 n_2) = 0 \qquad (3.15)$$

As $\sigma_1 \neq \sigma_2$, $(l_1 l_2 + m_1 m_2 + n_1 n_2)$ must be zero which in fact means that these two directions are perpendicular. In the same way it can be shown that the third direction is perpendicular to the first two directions.

Equation (3.14) can also be written as

$$\sigma^3 + I_1\sigma^2 + I_2\sigma + I_3 = 0 \qquad (3.16)$$

The roots of this equation are the principal stresses which are independent of the original choice of the coordinate system. Hence also the quantities I_1, I_2 and I_3 must be independent of this choice and hence are invariant. Worked out,

$$I_1 = \sigma_{xx} + \sigma_{yy} + \sigma_{zz} = \sigma_1 + \sigma_2 + \sigma_3 = 3P_d$$

$$I_2 = -(\sigma_{yy}\sigma_{zz} + \sigma_{zz}\sigma_{xx} + \sigma_{xx}\sigma_{yy}) + \sigma_{xy}^2 + \sigma_{xz}^2 + \sigma_{yz}^2$$

$$= -(\sigma_1\sigma_2 + \sigma_1\sigma_3 + \sigma_2\sigma_3)$$

$$I_3 = \sigma_{xx}\sigma_{yy}\sigma_{zz} + 2\sigma_{yz}\sigma_{xz}\sigma_{xy} - \sigma_{xx}\sigma_{yz}^2 - \sigma_{yy}\sigma_{zx}^2 - \sigma_{zz}\sigma_{xy}^2 = \sigma_1\sigma_2\sigma_3$$

These relations can be compared with the relations derived for the two-dimensional system (3.1)–(3.6).

Suppose now that the original coordinate system is chosen such that it coincides with the system of principal axes. Hence, the shear stresses σ_{xy}, σ_{xz} and σ_{yz} are zero. Again we consider a plane through the point O with direction cosines l, m and n such that $\sigma_1 > \sigma_2 > \sigma_3$. On this plane acts a stress **p** which can be resolved into a normal stress **σ** and a shear stress **τ**. From eqn (3.8) we find that $\sigma = l^2\sigma_1 + m^2\sigma_2 + n^2\sigma_3$.

When p is the magnitude of the stress **p** it follows that $p^2 = \sigma^2 + \tau^2$ but also $p^2 = p_x^2 + p_y^2 + p_z^2 = l^2\sigma_1^2 + m^2\sigma_2^2 + n^2\sigma_3^2$. Hence

$$\tau^2 = l^2\sigma_1^2 + m^2\sigma_2^2 + n^2\sigma_3^2 - \sigma^2 \qquad (3.17)$$

With $l^2 + m^2 + n^2 = 1$ we have three equations in l^2, m^2 and n^2. On elimination of m^2 and n^2 it is found that

$$\tau^2 + (\sigma_3 - \sigma)(\sigma_2 - \sigma) = l^2(\sigma_1 - \sigma_3)(\sigma_1 - \sigma_2)$$

and this can be rewritten into

$$\tau^2 + \{\sigma - \tfrac{1}{2}(\sigma_2 + \sigma_3)\}^2 = l^2(\sigma_1 - \sigma_3)(\sigma_1 - \sigma_2) + \tfrac{1}{4}(\sigma_2 - \sigma_3)^2 \qquad (3.18)$$

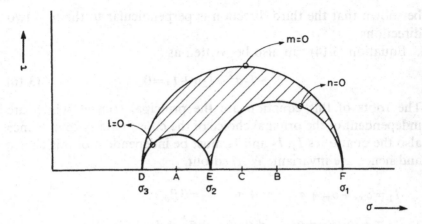

Fig. 3.5. Three-dimensional Mohr circles.

The possible combinations (σ, τ) for a fixed value of l are now plotted in a σ–τ diagram (see Fig. 3.5). From eqn (3.18) it follows that the corresponding points lie on a circle which has as its centre the point A on the σ-axis with the coordinate $\frac{1}{2}(\sigma_1 + \sigma_3)$ and whose radius r is given by

$$r^2 = l^2(\sigma_1 - \sigma_3)(\sigma_1 + \sigma_2) + \tfrac{1}{4}(\sigma_2 - \sigma_3)^2$$

When

$$l = 0, \quad r = \tfrac{1}{2}(\sigma_2 - \sigma_3) \quad \text{(Fig. 3.5: } r = \text{AE)}$$
$$l = 1, \quad r = (\sigma_1 - \tfrac{1}{2}\sigma_2 - \tfrac{1}{2}\sigma_3) \quad \text{(Fig. 3.5: } r = \text{AF)}$$

These are the Mohr circles already introduced in Section 3.2 for the case of a two-dimensional system. For intermediate values of l the corresponding radius is also intermediate. In the same way by elimination of l^2 and m^2 it is found that

$$\tau^2 + \{\sigma - \tfrac{1}{2}(\sigma_1 + \sigma_2)\}^2 = n^2(\sigma_3 - \sigma_2)(\sigma_3 - \sigma_1) + \tfrac{1}{4}(\sigma_1 - \sigma_2)^2 \quad (3.19)$$

The Mohr circles now have as centre the point $(\frac{1}{2}(\sigma_1 + \sigma_2), 0)$ which corresponds with point B in Fig. 3.5, while for

$$n = 0, \quad r = \tfrac{1}{2}(\sigma_1 - \sigma_2) \quad \text{(Fig. 3.5: } r = \text{BE)}$$
$$n = 1, \quad r = \tfrac{1}{2}(\sigma_1 + \sigma_2) - \sigma_3 \quad \text{(Fig. 3.5: } r = \text{BD)}$$

Finally, eliminating l^2 and n^2,

$$\tau^2 + \{\sigma - \tfrac{1}{2}(\sigma_1 + \sigma_3)\}^2 = m^2(\sigma_1 - \sigma_2)(\sigma_3 - \sigma_2) + \tfrac{1}{4}(\sigma_1 - \sigma_3)^2 \quad (3.20)$$

The centre of the Mohr circles now is at point C while for

$$m = 0, \quad r = \tfrac{1}{2}(\sigma_1 - \sigma_3) \quad \text{(Fig. 3.5: } r = \text{CD)}$$
$$m = 1, \quad r^2 = (\sigma_2 - \tfrac{1}{2}\sigma_3 - \tfrac{1}{2}\sigma_1)^2 \quad \text{(Fig. 3.5: } r = \text{CE)}$$

As eqns (3.18)–(3.20) all have to be satisfied at the same time, in Fig. 3.5 only combinations (σ, τ) in the hatched area are possible.

If $\sigma_1 = \sigma_2 = \sigma_3$ the radius r becomes zero. In this case no shear stress is working and we are dealing with a situation involving isotropic pressure only.

What has been said so far on the stress tensor applies to all (mechanical) structures and materials. When it is applied to a powder it must be realized that a stress as defined here (normal or shear stress) is the sum of all mechanical forces acting on a small surface through the powder divided by its surface area. (Of course we must exclude forces exerted on the particles by the surrounding fluid, such as buoyancy and viscous shear: see Chapter 4.) Since these forces can be very high at the contact points between the particles and will be zero at all those points where there is no contact, it will be clear that in order to come to a workable definition of the average local stress this surface area must be large compared to the average cross-section of a particle.

The subject matter further treated in this chapter specifically applies to powders.

3.4 THE YIELD LOCUS

When the circumstances at some special plane X are such that the shear stress acting on that plane reaches a critical value while the normal stress is not too high, the elasticity limit of the powder will be reached, the powder will yield and be sheared off along the plane X: the powder starts to flow. This critical shear stress

depends on the normal stress acting on the plane X and increases
when the normal stress increases, as indicated by the yield locus in
Fig. 3.6. If the powder is non-cohesive this yield locus will pass
through the origin of the (σ, τ) diagram. Fine powders, however,
generally are cohesive and in that case the yield locus will pass
through a point on the negative σ-axis, the so-called tensile stress
which is the maximum tensile stress that the powder can withstand.

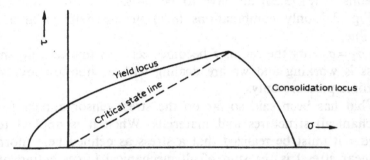

Fig. 3.6. Yield locus, consolidation locus, critical state line.

It follows that stress situations characterized by points above
the yield locus are not possible. Since the largest Mohr circle
determined by the principal stresses σ_1 and σ_3 (independent of σ_2
while $m = 0$, see Fig. 3.5) represents all maximum (σ, τ) combinations, it will be clear that this Mohr circle must lie entirely below
the yield locus and at most can touch this yield locus, in which case
the powder will yield. The point of contact between the Mohr
circle and the yield locus determines at the same time the value of
the direction cosines l and n ($m = 0$) of the plane X along which the
shearing off occurs. As long as the powder does not yield, only
small changes in porosity due to elastic deformation occur. When
the powder does yield this is accompanied by an increase of the
local porosity.

The foregoing holds as long as the normal stress does not exceed
a critical value σ_r. When this stress is exceeded the powder will not
be sheared off at increasing shear stress but will be compressed
(consolidated) so that its porosity will decrease. This is indicated

by the so-called consolidation locus. In this region the critical shear stress decreases with increasing normal stress until it becomes zero at a normal stress σ_m. This latter situation can only be reached when the applied stress situation is such that the three principal stresses become equal (isotropic pressure). Also in this region the porosity will change only little owing to elastic deformation as long as the consolidation locus is not reached. It follows that the critical normal stress indicates the boundary between yielding and consolidation. Hence, it must be expected that when at this normal stress the shear stress is increased the porosity will neither increase nor decrease when the critical shear stress is reached.

The yield locus and the consolidation locus according to the foregoing definitions both depend on the original porosity of the powder in such a way that they shift to higher values at decreasing porosity (see Fig. 3.7). Also the critical normal stress σ_r and the corresponding critical shear stress τ_r will increase at decreasing porosity. The line connecting the points (σ_r, τ_r) in the (σ, τ) diagram for the different porosities is called the critical state line which passes through the origin of the (σ, τ) diagram, also in the case of cohesive powders. According to Roscoe (1967) experiments have shown that this critical state line is

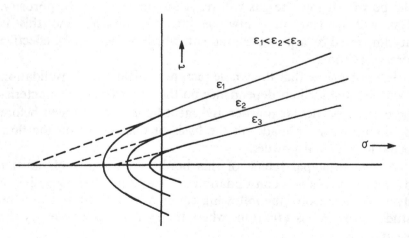

Fig. 3.7. Yield locus at increasing porosity.

a straight line through the origin. For certain powders this may be so but there is no theory available to support this statement.

If the stress situation is such that the powder is consolidated, the powder is in a stable state. This means that when the stress is intensified a new equilibrium is reached at a lower porosity. In the yield region the situation is more complicated. When the powder has reached a state in which it will yield along a plane X, the porosity will increase at X, but since the porosity is a space variable an increase at X also means an increase at $X + \Delta X$ and hence the porosity will yield also here, with a consequential further increase of the porosity. Thus, the region of increased porosity will rapidly extend. This goes on until the critical state line is reached where the porosity will not change any more. After some time the bulk of the powder will attain this state and then show stationary flow everywhere. At the same time the stresses must be relieved such that the radius of the critical Mohr circle decreases either by a decrease of the principal stress σ_1 or by an increase of the principal stress σ_3; how, exactly, will depend on the way in which the stresses from the outside are applied.

In chapter 7 it will be shown that for fine powders this depends also on the viscosity of the gas which is present inside the pores of the powder. When this viscosity is high and/or the permeability of the powder is low the gas will resist any increase of the porosity. Gas will tend to flow inwards into the powder and this is accompanied by a pressure gradient, which influences the effective stress situation.

It will be clear that the whole picture of yield and consolidation locus, as well as their dependence on the porosity, are characteristic to each particular powder and must be first determined before predictions can be made on the basis of this theory on the flow behaviour of that powder.

As the principal theme of this book is the study of powder dynamics and since consolidation is practically the opposite of dynamic behaviour, the following chapters will be confined to the study of low stress situations where the limit is represented by the yield locus.

In this chapter some applications of the foregoing theory will now be treated with the aid of two-dimensional situations, since these are more easily interpreted. To facilitate theoretical derivations the yield line will be approximated (for $\sigma > 0$) by a straight line, $\tau = \tau_0 + \sigma \tan \phi$ (the Coulomb criterion), where ϕ is called the angle of internal friction. For $-\sigma_c < \sigma < 0$ the yield locus is curved and passes through the point $(-\sigma_c, 0)$: see Fig. 3.8. The constant τ_0 is often called the cohesion constant. This name, however, is not appropriate since σ_c is the stress directly caused by cohesion between the particles.

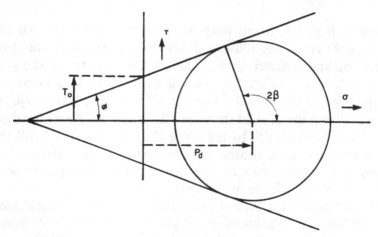

Fig. 3.8. Coulomb criterion.

3.5 THE ANGLE OF REPOSE

When a powder is dumped freely in a heap on a horizontal plane, this heap will show a slope which is characteristic of the powder in question and is called the angle of repose. For non-cohesive powders ($\tau_0 = 0$) the angle of repose is equal to the angle of internal friction and offers, therefore, the following simple method of determining ϕ.

We consider an imaginary heap with a slope $\beta < \phi$ (see Fig. 3.9). In this heap we consider a volume element ABCD with a thickness

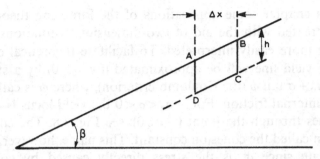

Fig. 3.9. Imaginary heap at rest with stresses in volume element near surface of heap.

Δx and a height h. If this heap is at rest the stresses on AB and CD must be equal and must neutralize each other, while the stress on DC makes equilibrium with the gravitational force. The total force on CD is $-\bar{\rho}gh(\Delta x)$ where $\bar{\rho}$ is the average bulk density of the powder. The normal force on CD is $-\bar{\rho}gh(\Delta x)\cos\beta$. As $CD = \Delta x/\cos\beta$ the normal stress on CD is $\bar{\rho}gh\cos^2\beta$ and the shear stress is $\bar{\rho}gh\cos\beta\sin\beta$. In the stress diagram (see Fig. 3.10) this stress situation is indicated by the point Z. The Mohr circle belonging to a point on CD, therefore, will pass through the point Z but will not touch the yield curve.

We will now do a mental experiment and put a vertical plate into the heap at each side of the volume element ABCD. By means of outward forces on these plates the stresses inside ABCD can be intensified (pressing the plates together) or relieved (pulling them apart). We do this in such a way that the heap comes to the verge of yielding (just stays at rest). The Mohr circle belonging to a point on CD will change but must still pass through the point Z since the stress on CD is determined only by gravitation. It appears now that two situations exist in which the powder is on the verge of yielding as indicated by the Mohr circles 1 (stress relief) and 2 (stress intensification). We consider Mohr circle 1. The point Q on this circle represents the shear plane through a point on CD. The angle QMB ($=2\phi$) now is twice the angle which this shear plane makes with the principal stress plane (see Section 3.2). The angle ZMB ($=2\gamma$) is twice the angle which this latter plane makes with

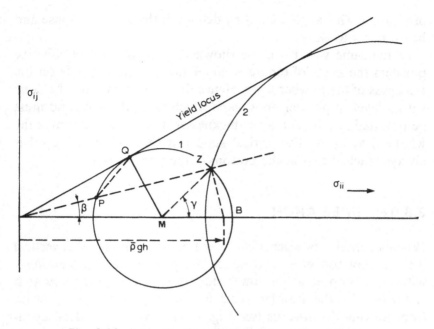

Fig. 3.10. Intensification of stresses by outward forces.

the plane through CD, that is with the slope of the heap. Hence
$(\theta - \gamma)$ is the angle which the shear plane makes with the slope of
the heap and this angle is equal to the angle QPZ. As PZ is the
direction of the slope of the heap, PQ must be the real direction
of the shear plane.

We will now increase the slope β in small steps (before each step
the vertical plates must be removed and after each step put into
the heap again while the critical stress situation is restored). The
Mohr circles 1 and 2 will now approach each other, which means
that at consecutive steps the critical situation just short of shearing
off is reached at ever smaller outward forces. At the same time the
points Z and P both approach the point Q. When the angle β becomes
equal to ϕ the two Mohr circles 1 and 2 coincide and without
any outward force this Mohr circle through Z will touch the yield
curve. The direction PQ has become equal to the slope of the heap
and we have reached the situation where the heap is on the verge

of yielding. This angle then is by definition the angle of repose and hence equal to ϕ.

In the same way it can be shown that in the case of cohesive powders the angle of repose is larger than ϕ and depends on the thickness of the powder layer. Hence the experiment must be done with a layer of powder on a plane which is slowly tilted and must be repeated at different layer thicknesses in order to determine the whole yield locus. The critical state just short of shearing off is always reached first at the bottom of the powder layer.

3.6 THE FREE ARCH

When a cohesive powder is contained between two parallel vertical walls at not too great a distance the powder can be stationary without a support at the lower side. In fact a so-called free arch is formed by the powder. This free arch prevents the powder from flowing downwards (see Fig. 3.11). Also in a vertical cylin-

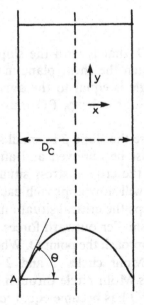

Fig. 3.11. Free arch.

drical vessel of not too large a diameter a free arch can be formed.

The stress situation in the powder is described by the differential equation

$$\nabla \cdot \boldsymbol{\sigma} - \bar{\rho}\mathbf{g} = 0 \qquad (3.21)$$

In the two-dimensional case (x–y coordinates) this equation can be resolved into two scalar equations:

$$\frac{\partial}{\partial x}\sigma_{xx} + \frac{\partial}{\partial y}\sigma_{xy} = 0 \qquad (3.22)$$

$$\frac{\partial}{\partial y}\sigma_{yy} + \frac{\partial}{\partial x}\sigma_{xy} + \bar{\rho}g = 0 \qquad (3.23)$$

When the powder is at rest and has formed a free arch, the whole mass of powder is carried by the two walls by means of wall friction. As this is independent of the total mass of powder kept between the two walls, it follows easily that—apart from a transition zone at the top—the stress situation inside the powder must be independent of the vertical y-coordinate. Hence

$$\frac{\partial}{\partial y}\sigma_{xy} = 0 \quad \text{and} \quad \frac{\partial}{\partial y}\sigma_{yy} = 0$$

From eqns (3.22) and (3.23) it then follows that

$$\frac{\partial}{\partial x}\sigma_{xx} = 0 \quad \text{or} \quad \sigma_{xx} = q$$

and

$$\sigma_{xy} = -\bar{\rho}gx + \text{const.}$$

where q is a constant still unknown.

With $x=0$ in the central plane between the two walls and from the consideration that the stress profile must be symmetrical in relation to this central plane, it follows that

$$\sigma_{xy} = -\bar{\rho}gx \qquad (3.24)$$

The Mohr circle pertaining to a point $x = -l$ of the free arch goes through the point $(q, \bar{\rho}gl)$ in the (σ, τ) diagram. Since the free arch is a free surface, the normal stress and the shear stress at this surface must disappear. Hence, the Mohr circle also passes through the origin of the (σ, τ) diagram and this means that this Mohr circle is fixed (see circle (1) in Fig. 3.12). The largest possible

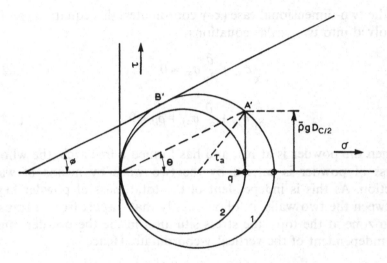

Fig. 3.12. Mohr circle of free arch.

Mohr circle pertains to the point A in Fig. 3.11. When this Mohr circle touches the yield curve, it gives the maximum possible span D_c of the free arch. When r_a is the radius of this Mohr circle, the following relation holds:

$$r_a = \tau_0 \cos \phi + r_a \sin \phi$$

or

$$r_a = \frac{\tau_0 \cos \phi}{1 - \sin \phi} \quad \text{(independent of } q\text{)}$$

From Fig. 3.12 it follows that also

$$r_a = B^2/8q + q/2 \quad \text{with } B = \bar{\rho}gD_c \tag{3.25}$$

It can now be derived that

$$D_c = \frac{2}{\bar{\rho}g}\sqrt{2qr_a - q^2} \qquad (3.26)$$

and

$$q = r_a + \sqrt{r_a^2 - B^2/4}$$

When we put $q = nr_a$ this gives

$$D_c = \frac{2r_a}{\bar{\rho}g}\sqrt{2n - n^2}$$

from which it follows that $n < 2$. The above result can also be written as

$$n = 1 + \sqrt{1 - \left(\frac{B}{2r_a}\right)^2}$$

from which it follows that $n > 1$. Hence $1 < n < 2$. As probably $1 < n < 1.5$ it follows from Table 3.1 that it can be expected that

$$D_c \approx 1.8\, r_a/\bar{\rho}g \qquad (3.27)$$

Note also that free arches can only be formed by cohesive powders ($\tau_0 > 0$).

So far it has been tacitly assumed that the friction of the powder at the wall is strong enough to prevent the whole mass of the

TABLE 3.1
The Function $f(n) = \sqrt{2n - n^2}$

n	$\sqrt{2n - n^2}$
1·0	1·000
1·1	0·995
1·2	0·980
1·3	0·954
1·4	0·916
1·5	0·866
1·8	0·600
2·0	0·000

powder from slipping down by gravity. Whether this occurs will depend on the yield stress at the wall for which there is a critical yield curve as well. This critical yield curve is independent of the yield of the powder itself and is determined by the properties of both the wall and the powder. Also at the wall there can be some kind of cohesion between the wall and the powder which now is called adhesion. The slope of the wall yield curve is the angle of wall friction which to a large extent is determined by the wall roughness.

The foregoing means that in order to prevent the powder mass from sliding downwards the point A' in Fig. 3.12 must also be below the wall yield curve. Now, however, this condition is sufficient as there is no free choice for the slope of the shear plane, this being fixed by the presence of the wall. Hence, the critical Mohr circle may cut the wall yield curve as long as the point A' is below this curve.

The shape of the arch is determined by the angle θ which the x-axis makes with the principal axis I and the variation of this angle with x. It must be realized that the direction of this axis varies with x as well and is everywhere equal to the local direction of the tangent to the arch.

Hence, along the free arch:

$$\frac{dy}{dx} = \tan\theta = \bar{\rho}gx/q$$

It follows that $y = \bar{\rho}gx^2/2q$. Where the arch reaches the wall, $\tan\theta = \bar{\rho}gD/2q$. If $q = r_a$ it follows that $\theta = 45°$.

3.7 INCIPIENT FLUIDIZATION

When a powder is contained in a vertical cylindrical vessel (or between two parallel vertical walls) closed at the bottom by a horizontal porous plate that is permeable to a gas but impermeable to the powder particles, and open at the top, such a vessel is called a fluidization bed. When a gas is supplied from below through the porous plate, the gas flow exercises an upward drag force on the particles. The differential vector equation which describes the stress in the powder mass must now be extended with the pressure gradient:

$$\nabla \cdot \boldsymbol{\sigma} + \nabla p - \bar{\rho}\mathbf{g} = \mathbf{0} \qquad (3.28)$$

When elaborated in cylindrical coordinates this gives, for the radial direction,

$$\frac{1}{r}\frac{\partial}{\partial r}(r\sigma_{rr}) - \frac{\sigma_{\theta\theta}}{r} + \frac{\partial}{\partial y}\sigma_{ry} = 0 \qquad (3.29)$$

and for the vertical direction,

$$\frac{\partial}{\partial y}\sigma_{yy} + \frac{1}{r}\frac{\partial}{\partial r}(r\sigma_{ry}) + \bar{\rho}g + \frac{dp}{dy} = 0 \qquad (3.30)$$

In this latter equation $\bar{\rho}g$ has a constant value while dp/dy depends only on the gas flow rate v_{co}. Note that dp/dy is negative. Of course, there is also a scalar differential equation for the tangential direction θ, but as we assume cylindrical geometry and no tangential movement this third equation is not of interest to us.

Since we deal only with two equations (3.29 and 3.30) but have four unknowns (σ_{rr}, σ_{yy}, σ_{ry} and $\sigma_{\theta\theta}$), the problem cannot be solved without further assumptions. One assumption is again that the stress situation will be invariant with respect to the vertical y-coordinate. This gives, after integration,

$$\sigma_{rr} = \frac{1}{r}\int_0^r \sigma_{\theta\theta}\, dr$$

and

$$\sigma_{ry} = -\frac{Ar}{2} + \frac{C_1}{r}$$

with $A = \bar{\rho}g + dp/dy$. Because at $r \to 0$ the stresses must remain finite it follows that $C_1 = 0$. Also for $r \to 0$ $\sigma_{\theta\theta}$ must $\to 0$. Hence, it seems logical to put $\sigma_{\theta\theta} = C_2 r$ which gives

$$\sigma_{rr} = \frac{C_2}{2}r \quad \text{and} \quad \sigma_{ry} = -\frac{Ar}{2} \qquad (3.31)$$

As long as A is positive (at low values of $-dp/dy$) the powder structure will remain the same. When at increasing gas flow rate A becomes zero the gas flow rate at incipient fluidization is reached. At further increase of the flow rate the powder will try to expand but is prevented from this by the internal friction. When A has reached a (negative) value strong enough to overcome this friction, the bed will indeed shear off and expand, so that the pores between the particles increase in size with the consequence that the drag force on the particles again reaches equilibrium with the weight of the particles. Hence, A becomes zero again. The powder bed is then fluidized. The total pressure drop across the bed (with bed height H) is given by

$$\Delta p = -\int_0^H \frac{dp}{dy} \cdot dy \qquad (3.32)$$

At low gas flow rate, when A is positive and H is constant, Δp increases linearly with the gas flow rate v_{co} (at least in the case of fine powders where the Reynolds number $\mathrm{Re} = \rho_c v d_p / \mu \ll 1$). When the flow rate is increased beyond the gas flow rate at incipient fluidization, Δp will first increase further as long as the bed is prevented from expansion by the internal friction, but will return to a constant value $\Delta p_F = W/O$ (in which W is the total weight of the powder mass and O the cross-sectional area of the bed) when expansion has started (see Fig. 3.13).

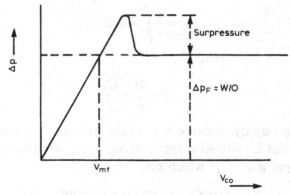

Fig. 3.13. Δp Versus v_{co} diagram.

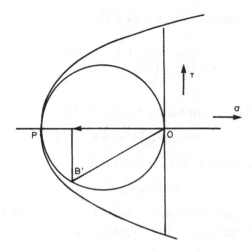

Fig. 3.14. Critical Mohr circle just before expansion.

The stress situation just before expansion will now be analysed. Figure 3.14 presents the relevant (σ, τ) diagram with the yield locus. As already mentioned, the critical situation just before bed expansion is characterized by a negative value of the parameter A and so the critical Mohr circle will lie entirely to the left of the τ-axis. As the bed surface is a free surface it can be concluded again that the origin of the (σ, τ) diagram must be a point of this Mohr circle, while this circle also must touch the yield locus. Although this yield locus to the left is not exactly known, it is fairly sure that σ_c will not be much larger than τ_0. In that case the largest possible Mohr circle will touch the yield curve in the point $(-\sigma_c, 0)$ (see Fig. 3.14).

If R is the bed radius, the point given by $\sigma = C_2/R$ and $\tau = -AR/2$ must be a point of this Mohr circle. Hence, when r_a is the radius of this Mohr circle, we have

$$2r_a = \sigma_c$$

and

$$r_a^2 = (\sigma - r_a)^2 + \tau^2$$

After substitution of σ and τ and elaboration, one obtains

$$C_2 \frac{R\sigma_c}{2} - \left(\frac{C_2 R}{2}\right)^2 - \frac{A^2 R^2}{4} = 0$$

With $n = C_2 R/\sigma_c$ it can now be derived that

$$n = 1 + \sqrt{[1 - (AR/\sigma_c)^2]}$$

from which it follows that $1 < n < 2$ and

$$-AR/\sigma_c = \sqrt{(2n - n^2)}.$$

The 'surpressure' $\Delta p'$ across the bed necessary to overcome the stresses in the bed ($\Delta p' = \Delta p - W/O$) is related to A by

$$\Delta p' = -AH = \frac{H\sigma_c}{R}\sqrt{2n - n^2}$$

Assuming that $1 < n < 1{\cdot}5$, one finds that $\sqrt{(2n - n^2)} \approx 0{\cdot}9$. See again Table 3.1. Hence, by experimental measurement of $\Delta p'$ a fair estimate can be made of the tension cutoff:

$$\Delta p' \approx 0{\cdot}9 \frac{H\sigma_c}{R}$$

REFERENCES

Jaeger, J.C. (1969). *Elasticity, Fracture and Flow.* Science Paperbacks, Methuen, London.

Roscoe, K.H. (1967). Behandlung bodenmechanischer Probleme auf der Grundlage neuerer Forschungsergebnisse Bergbauwissenschaften, **14**, 464.

4

Theoretical Derivation of Interparticle Forces†

NOTATION

A	Hamaker constant (J)
C_{ij}	Constant in Lennard-Jones potential for substances i and j (J m^6)
d	Particle diameter, or diameter of the spring wire (m)
D	Equivalent diameter of contacting particles (m)
e	Free space between windings of spring (—)
E	Elasticity modulus of powder (N m^{-2})
E_m	Elasticity modulus of spring material (N m^{-2})
E_{pc}	Elasticity modulus of particle contact (N m^{-2})
E_s	Elasticity modulus of spiral spring (N m^{-2})
F	Force between particles (N)
F_c	Cohesion force (N)
G	Gas adsorption function (—)
h	Flattening of particles (m)
k	Boltzmann constant (J K^{-1})

†The analysis in this chapter is a further elaboration and explanation of the paper by Cottaar and Rietema (1986).

K	$= (1 - v^2)/Y(N^{-1}\,m^2)$
L	Distance between particle centre and surface of the plane (m)
n	Molecule density (m^{-3})
N_a	Density of adsorbed gas molecules (m^{-2})
N_{oa}	Maximum density of adsorbed gas molecules (m^{-2})
r	Distance between molecules (m)
r_{ij}	Molecule parameter (see eqn (4.5)) (m)
R	Radius of particle, or radius of spiral spring (m)
s	Parameter indicated in Fig. 4.4 (m)
T	Temperature (K)
U_m	Total molecular potential of system (J)
$U_c(y)$	Binding energy between two planes (J)
V	Volume of particle (m^3)
V_{ij}	Intermolecular potential for substances i and j(J)
Y	Young's modulus of elasticity $(N\,m^{-2})$
z	Smallest distance between the surfaces of the particle and the plane (m)
α	Ratio of constants for molecular solid–solid interaction and gas adsorption (—)
δ	Fraction of adsorbed gas density (—)
ε_c	Dimensionless interaction parameter (—)
ζ	Dimensionless distance (—)
η	Dimensionless flattening (—)
θ_c	Dimensionless density of adsorbed gas molecules (—)
v	Poisson's ratio (—)
ρ_d	Density of solid particles $(kg\,m^{-3})$
σ_c	Cohesion constant $(N\,m^{-2})$
σ_n	Normal stress $(N\,m^{-2})$
τ	Ratio of adsorption energy to thermal energy (—)
ϕ	Dimensionless external force applied to the particle–plane system (—)
χ	Dimensionless interaction energy (—)
ψ	Dimensionless elastic deformation energy (—)
ω	$= \zeta - \eta$ (—)

4.1 INTRODUCTION

The so-called interparticle forces comprise capillary forces, electrostatic forces and Van der Waals forces.

Capillary forces are caused by moisture which condenses from the surrounding gas on the surface of the particles and then forms liquid bridges between neighbouring particles so that the action of the surface tension gives rise to attractive forces. These forces can be very strong (many times the weight of the individual particles) when the vapour pressure of the surrounding gas is close to the saturation pressure. In operations such as drying of the solids they may seriously hamper the handling of the powder, especially at the start. In most powder operations, however, vapour pressures are so low that these capillary forces can be neglected.

The second type of interparticle force is electrostatic. When non-conducting particles are contacting surfaces of dissimilar material or sliding along such surfaces, this is generally accompanied by exchange of electrons in the surface layer. This causes the particles to be electrically charged (tribo-electricity) either positively or negatively. The direction of exchange depends on the nature of the contacting surfaces. The overall effect generally is that all particles are charged similarly. Hence according to Coulomb's law the interparticle forces caused by this phenomenon are repelling and not attracting.

The general opinion is that electrostatic forces are insignificant compared to the third class of interparticle forces, viz. Van der Waals forces. These operate between molecules at very short range. When these are integrated over all pairs of molecules around the point where the particles are in contact, the resulting interparticle forces may reach a considerable magnitude. This chapter will be devoted to this third class of interparticle forces. Special attention will be paid to (1) the effect of particle deformation at the place of contact between two neighbouring particles as a consequence of the attractive force; and to (2) the effect of gas adsorption, which increases the attractive force in most cases.

4.2 THE HAMAKER THEORY

Many investigators (Boehme *et al.*, 1962; Krupp, 1967; Johnson *et al.*, 1971; Dahneke, 1972; Derjaguin *et al.*, 1975; Tabor, 1977; Pollock, 1978) have occupied themselves, both theoretically and experimentally, with the attraction forces between solid particles on the basis of the Van der Waals forces. These forces were first derived theoretically by London (1937). According to his theory the interaction energy between two molecules is given by

$$V_{ss} = -\frac{C_{ss}}{r^6} \qquad (4.1)$$

in which C_{ss} is the London–Van der Waals constant, measured in $J\,m^6$ which depends on the nature of the interacting molecules, and r is the distance between these molecules. Although these forces operate only at extremely short range, they can reach values many times the weight of the particles when integrated over all molecules in two contacting particles. This was first shown by Bradley (1932) and Hamaker (1937). Integration is carried out according to

$$U_m = \int_{V_1} dn_1 \int_{V_2} V_{ss}\, dn_2 \qquad (4.2)$$

in which n_1, n_2, dV_1 and dV_2 designate molecule densities and volume elements of the two interacting particles respectively. For two perfectly spherical and rigid particles with diameters d_1 and d_2 this integration leads to

$$U_m = -\frac{A}{12z} \frac{d_1 d_2}{(d_1 + d_2)} \qquad (4.3)$$

where A is the so-called Hamaker constant, $A = \pi^2 n_1 n_2 C_{ss}$, while z is the shortest distance between the particles. Equation (4.3) holds only as long as $z \ll D = d_1 d_2/(d_1 + d_2)$. This, however, is not a practical limitation.

The attractive force which now follows from eqn (4.3) is found by differentiation with respect to z:

$$F = \frac{\partial}{\partial z}(U_m) = \frac{A}{12z^2}D \tag{4.4}$$

It has always been a point of discussion what value is to be filled in for z. It was realized that z cannot become zero as at still shorter distances strong repulsive forces are generated by the interaction of the electron orbits of the molecules. This is indicated by the potential according to Lennard-Jones (1937) which in fact is an extension of the London–Van der Waals potential and which runs:

$$V_{ss} = C_{ss}\left(-\frac{1}{r^6} + \frac{r_{ss}^6}{2r^{12}}\right) \tag{4.5}$$

in which r_{ss} is a characteristic parameter of the interacting molecules. As the repelling force is an even steeper function of the distance than the attraction, it follows that at equilibrium (net force $=0$) the shortest distance z_0 between the particles cannot vary very much. At increasing z above this value the net force becomes attractive and reaches a maximum at $z = z_m$. At this distance the repelling force has practically become zero and hence the cohesion force F_c can be calculated from eqn (4.4) with $z = z_m$ (for which most investigators have assumed a value of $4 \cdot 0 \times 10^{-10}$ m).

A better way, of course, is to integrate eqn (4.5) according to eqn (4.2). This leads to

$$U_m = \frac{AD}{12z}\left[-1 + \frac{1}{420}\left(\frac{r_{ss}}{z}\right)^6\right] \tag{4.6}$$

or in dimensionless form with $\zeta = z/r_{ss}$:

$$\chi = \frac{12U_m r_{ss}}{AD} = -\frac{1}{\zeta} + \frac{1}{420}\frac{1}{\zeta^7} \tag{4.6a}$$

This function χ is plotted in Fig. 4.1. At the point 0 the potential is at a minimum, hence the net force between the particles is

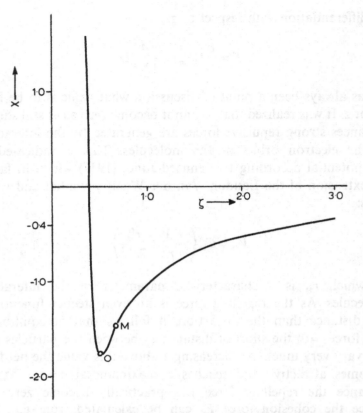

Fig. 4.1. Dimensionless interaction potential between two neighbouring particles based on the Lennard-Jones potential between molecules.

zero. When the distance between the particles is increased the force becomes attractive to reach a maximum at the point M where the gradient of the potential is maximum. The force F is given by

$$F = \frac{\partial U_m}{\partial z} = \frac{AD}{12z^2}\left[1 - \frac{1}{60}\left(\frac{r_{ss}}{z}\right)^6\right] \qquad (4.7)$$

or in dimensionless form:

$$\phi = \frac{12Fr_{ss}^2}{AD} = \frac{1}{\zeta^2}\left(1 - \frac{1}{60\zeta^6}\right) \qquad (4.7a)$$

From this relation it follows that $F = 0$ or $\phi = 0$ at $\zeta = 0.505$. The maximum F_c is found when $d\phi/d\zeta = 0$ at a value $\zeta = 0.6368$ which gives

$$F_c = 1.85 \frac{AD}{12 r_{ss}^2} \tag{4.8}$$

An average value of r_{ss} is 3.5×10^{-10}m so that z_m becomes 2.23×10^{-10} m (compare this with the above-mentioned value $z_m = 4 \times 10^{-10}$ m). For most solids an average value of $A = 10^{-19}$ J has been calculated (for a survey see Visser (1972)).

With the above value of r_{ss} and with $d_1 = d_2 = 50$ μm it is found that the cohesion force between the particles F_c is 31×10^{-7} J. Comparison of this cohesion force with the weight of the particles gives $F_c/F_{gravity} \simeq 1900$ for glass beads ($\rho_d = 2600 \text{ kg m}^{-3}$) and $F_c/F_{gravity} = 3600$ for potato starch ($\rho = 1400 \text{ kg m}^{-3}$).

These seem extremely high values. Indeed for potato starch whose discrete particles have an extremely smooth surface in comparison with most other powders (see Fig. 1.1), a very high cohesion can be observed which makes handling of this powder as such practically impossible. Most powders, however, have a rather rough surface with many protuberances—generally called asperities—with radii of curvature often not larger than 0.1 μm (Donsi & Massimilla, 1973; Massimilla & Donsi, 1976). This means that the effective contact area between particles is determined by one, two or at most three asperities. To calculate the cohesion force per asperity, not the diameter of the particles but the diameter of the asperities should be inserted in eqn (4.8) which, of course, strongly reduces the derived cohesion force. The effective average cohesion force, therefore, depends strongly on the surface structure of the particles, i.e. the size distribution of the asperities and their surface density.

All these considerations make it very unlikely that the average cohesion force between particles of a powder can ever be predicted with reasonable accuracy. Nevertheless, a qualitative insight in the mechanism of cohesion will be of great help in understanding powder dynamics.

4.3 EFFECT OF PARTICLE DEFORMATION

In the analysis so far we have supposed that the particles are
entirely rigid and purely spherical. As shown above the sphericity
is of importance only at and around the contact area of the particle
or its asperities. Solids, however, are generally not rigid but more
or less deformable when pressed together, e.g. as a result of
interparticle forces. The deformation thus obtained will be an
elastic deformation when the compressing forces are not too
strong. This means that the process of deformation is reversible
and that the original shape of the particles will be restored when
the compressing force is removed.

When, however, the compressing forces exceed a certain limit
(the elasticity limit), the elastic deformation is followed by plastic
deformation which is not reversible and remains when the compressing force is reduced. It is, besides, a process that continues as
long as the compressing forces are maintained. It is believed that in
the dynamics of expanded powders as they occur in fluidization
and many powder-handling operations, plastic deformation can be
neglected and that, hence, we have to account for elastic deformation only.

Elastic deformation causes the particles (or asperities) to be
flattened at the contact area with the result that this area is
enlarged. In Fig. 4.2 this is shown for the case of an elastic

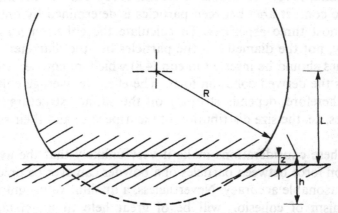

Fig. 4.2. The geometry of a flattened particle adhering to a plane.

spherical particle interacting with a flat rigid plane. Dahneke (1972) argued that this effect would enlarge the cohesion force between the particles. In his analysis, however, he assumed that z_m would be constant and, hence, he neglected the repelling force according to Lennard-Jones. He only took into account the repelling force which is due to the elastic deformation of elastic bodies when pressed together. According to Hertz (1895) this force is equal to

$$F_{rep} = \frac{4}{3K} \sqrt{Dh^3}$$

in which $K = (1 - v^2)/Y$ where v is Poisson's ratio and Y is Young's modulus of the solid. Again

$$D = \frac{d_1 d_2}{d_1 + d_2}$$

The distance L between the particle centre and the rigid plane now depends on two variables, viz. the distance z between the two surfaces and the flattening h of the particle:

$$L = R + z - h$$

Hence the system has one degree of freedom left. Again the force F is found from the interaction potential U by

$$F = \frac{dU}{dL}$$

At equilibrium with an external force at constant distance L the potential will be at a minimum and therefore will not vary with z or h. This means that at equilibrium

$$\left(\frac{dU}{dz}\right)_L = 0 \quad \text{and} \quad \left(\frac{dU}{dh}\right)_L = 0$$

or

$$\left(\frac{\mathrm{d}U}{\mathrm{d}z}\right)_L = \frac{\partial U}{\partial z} + \frac{\partial U}{\partial h} \cdot \frac{\mathrm{d}h}{\mathrm{d}z} = \frac{\partial U}{\partial z} + \frac{\partial U}{\partial h} = 0 \qquad (4.9)$$

since at constant L we have $\mathrm{d}h/\mathrm{d}z = 1$.

Integration of eqn (4.5) over the volume of the contacting particles now results in

$$U_\mathrm{m} = \frac{AD}{12z}\left[-\left(1 + \frac{h}{z}\right) + \frac{1}{60}\left(\frac{1}{7} + \frac{h}{z}\right)\left(\frac{r_{ss}}{z}\right)^6 \right] \qquad (4.10)$$

The energy due to the elastic deformation of a sphere must be accounted for too. After integration of the Hertz force it is given by

$$U_\mathrm{H} = \frac{8}{15}\frac{\sqrt{D}}{K}h^{2.5}$$

Hence, the total potential energy U is found to be

$$U = \frac{AD}{12z}\left[-\left(1 + \frac{h}{z}\right) + \frac{1}{60}\left(\frac{1}{7} + \frac{h}{z}\right)\left(\frac{r_{ss}}{z}\right)^6 \right] + \frac{8}{15}\frac{\sqrt{D}}{K}h^{2.5}$$

With $\eta = h/r_{ss}$ and $\psi = 32r_{ss}^{3.5}/5AK\sqrt{D}$ this can be written in dimensionless form as

$$\chi = -\left(\frac{1}{\zeta} + \frac{\eta}{\zeta^2}\right) + \frac{1}{60}\left(\frac{1}{7\zeta^7} + \frac{\eta}{\zeta^8}\right) + \psi\eta^{2.5} \qquad (4.12)$$

After rewriting eqn (4.9) in dimensionless form as

$$\frac{\partial \chi}{\partial \zeta} + \frac{\partial \chi}{\partial \eta} = 0 \qquad (4.9a)$$

it can be elaborated with the help of eqn (4.12), which finally results in

$$\sqrt{\eta} = \frac{\zeta^{-6} - 15}{\frac{75}{4}\psi\zeta^3} \qquad (4.13)$$

For each value of $\zeta < 0.6368$ η can now be calculated from eqn (4.13). Hence, also χ can be calculated. For $\zeta > 0.6368$ $\sqrt{\eta}$ becomes negative and then it is assumed that $\eta = 0$.

Although χ has been calculated as a function of ζ, it will be plotted versus $\omega = \zeta - \eta$, since ω indicates the real dimensionless distance between the particles. It is the increase of this distance that requires work to be done which results in an increase of the potential energy of the particle system. Hence, the attractive force between the particles is found by differentiation of χ with respect to ω.

In Fig. 4.3 χ is plotted for two different values of ψ together

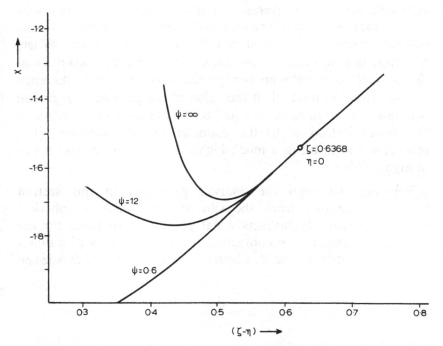

Fig. 4.3. Plot of dimensionless interaction potential between deformed particles versus $\omega = \zeta - \eta$ ($\psi = 2$, $\psi = 10$). Comparison with rigid particles ($\eta = 0$).

with the result for rigid particles ($\eta = 0$ for all values of ζ). It is notable that all three curves merge into the same curve for $\omega \geqslant 0.6368$ where $\eta = 0$, while at $\omega = 0.6368$ all curves reach the same steepest slope, indicating that at this point the maximum attractive force (or cohesion) is reached and that it is the same for all three curves. Hence, the cohesion is given by $\phi = 1.85$ irrespective of the hardness of the solid. In a somewhat different analysis this same result was derived by Derjaguin *et al.* (1975) and by Johnson *et al.* (1971).

Nevertheless, this result is in strong contradiction to experimental evidence on organic powders such as a variety of polymers and foodstuffs. These powders are observed to be much more cohesive than crystalline and mineral powders. This is the more surprising since for organic powders the Hamaker constants are on the average lower than for mineral powders. An explanation based on the undeniably lower hardness of organic powders seems to be ruled out by the above theory. When, however, this hardness is low enough to cause the asperities at the contact place between two particles to be entirely flattened by the attractive force, then the value of the parameter D in eqn (4.6) must be assumed to be equal to the diameter of the particles themselves instead of to the diameter of the asperities. This, indeed, would result in a much higher attractive force and hence in higher cohesion.

> Remark: Although the analysis presented in this section started from the situation in which a spherical particle contacts a flat plane, it also holds for the case of two contacting spherical particles with diameters d_1 and d_2 respectively, in which case D is taken as

$$D = \frac{d_1 d_2}{d_1 + d_2}$$

4.4 EFFECT OF GAS ADSORPTION

Physical gas adsorption to solid surfaces is caused by the same Van der Waals forces that cause the cohesion between solid particles. When gas molecules are adsorbed to the surface of solid particles the adsorption energy is released. The amount released is highest in the fissure around the contacting point of two particles as here the gas molecules are adsorbed to two surfaces. When the particles are torn apart this extra adsorption energy must be provided by the external force, which hence must be larger than in the absence of adsorption.

The foregoing means that to the potential energy U of the (particle–particle or particle–flat plane) system should now be added the contribution of gas adsorption. To calculate this contribution two things must be known, viz.

(a) the binding energy $U_c(y)$ of a single molecule which is bound between two planes a distance y apart; and
(b) the number $N_c(y)$ of molecules adsorbed per unit surface area, which itself is influenced by the binding energy $U_c(y)$.

The total energy due to adsorbed gas now equals

$$U_g(h, z) = \int_0^\infty N_c(y) \cdot U_c(y)\, 2\pi s\, ds \qquad (4.14)$$

in which s has the meanings indicated in Fig. 4.4. From this figure

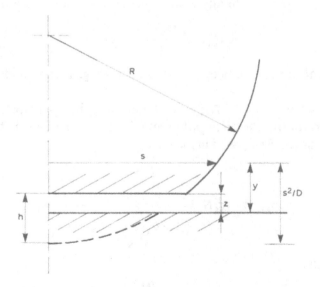

Fig. 4.4. Sketch of the geometry for the derivation of the energy due to gas adsorption.

it follows that

$$y = z \qquad \text{for } s < \sqrt{2RH}$$
$$y = z + s^2/2R - h \qquad \text{for } s \geqslant \sqrt{2RH}$$

Hence, for $s < \sqrt{2RH}$ both $U_c(y)$ and $N_c(y)$ are independent of s and equal $U_c(z)$ and $N_c(z)$ respectively.

For $s \geqslant \sqrt{2Rh}$ it is found that $s\,ds = R\,dy$ so that

$$U_g(h, z) = 2\pi Rh \cdot U_c(z) \cdot N_c(z) + 2\pi R \int_z^\infty U_c(y) \cdot N_c(y) \cdot dy \quad (4.15)$$

The exact derivation of both $U_c(y)$ and $N_c(y)$ is given in the appendix. The derivation of $U_c(y)$ is again based on the interaction energy between an individual gas molecule and an individual molecule of the solid as given by the Lennard–Jones potential,

$$V_{gs} = C_{gs}\left(-\frac{1}{r^6} + \frac{r_{gs}^6}{2r^{12}}\right) \quad (4.16)$$

With $\zeta_1 = y/r_{gs}$ this finally results in the dimensionless parameter

$$\varepsilon_c(\zeta_1) = \frac{12U_c r_{ss}}{AD} = \frac{-3}{\zeta_1^3} + \frac{8}{\zeta_1^9} \quad (4.17)$$

In the above the indices g and s refer to gas and solid respectively.

The derivation of $N_c(y)$ is based on the principle that at equilibrium the chemical potential of the adsorbed gas is the same as that of the free gas. This leads to

$$\theta_c(\zeta_1) = r_{gg}^2 N_c = \frac{\delta\left\{\left(\frac{-4}{\sqrt{2}}\varepsilon_c(\zeta_1) - 1\right)\tau\right\}}{1 + \delta\left\{\left(\frac{-4}{\sqrt{2}}\varepsilon_c(\zeta_1) - 1\right)\tau\right\}} \quad (4.18)$$

in which

$$\delta = \frac{N_a}{N_{oa} - N_a}$$

where N_a is the number of adsorbed gas molecules and N_{oa} the total number of available adsorption sites. Further $\tau = -U_a/kT$ in which U_a is the binding energy of a single gas molecule adsorbed to a solid plane, T is the temperature and k is the Boltzmann constant. τ can be further elaborated into

$$\tau = \frac{4\sqrt{2}}{9}\frac{\sqrt{AC_{gg}}}{r_{gs}^3 kT}$$

If we now define the function $G(\zeta) = \varepsilon_c(\zeta) \cdot \theta_c(\zeta)$ then the total potential energy can be written in dimensionless form:

$$\chi_t = -\left(\frac{1}{\zeta} + \frac{\eta}{\zeta^2}\right) + \frac{1}{60}\left(\frac{1}{7\zeta^7} + \frac{\eta}{\zeta^8}\right) + \psi\eta^{2.5}$$

$$+ \alpha\left[G(\zeta)\eta + \frac{r_{gs}}{r_{ss}}\int_\zeta^\infty G(\zeta_1)\mathrm{d}\zeta_1\right]\left(\frac{r_{ss}}{r_{gs}}\right)^2 \tag{4.19}$$

in which $\alpha = 48\pi U_a/AV_2$.

In the remainder of this chapter it will be assumed that the values of r_{ss}, r_{gs} and r_{gg} are equal. (This is not always justified as, e.g., in the case of neon r_{gg} is probably significantly smaller than r_{ss} for most solids.)

When we now apply eqn (4.9) again we find for the equilibrium relation between η and ζ that

$$\sqrt{\eta} = \frac{2}{5\psi}\left\{-\alpha\frac{\mathrm{d}}{\mathrm{d}\zeta}G + \frac{2}{15}\zeta^{-9} - 2\zeta^{-3}\right\} \tag{4.20}$$

Also in this case it is assumed that $\eta = 0$ whenever, according to eqn (4.20), $\sqrt{\eta}$ should be negative.

The dimensionless force ϕ can be derived by differentiating χ_t with respect to ω ($\omega = \zeta - \eta$). In the case that $\eta = 0$, ϕ can be derived directly from

$$\phi = \frac{\partial}{\partial\zeta}(\chi_t) = \frac{1}{\zeta^2}\left(1 - \frac{1}{60\zeta^6}\right) - \alpha G \tag{4.21}$$

while the value of ζ at which ϕ is a maximum can be derived directly from

$$\frac{d\phi}{d\zeta} = \frac{2}{\zeta^3} - \frac{2}{15}\frac{1}{\zeta^9} - \alpha\frac{dG}{d\zeta} = 0 \qquad (4.22)$$

It will be clear that the effect of gas adsorption will be to make the function χ_t much more complicated. In order to understand this effect we must first consider the function G more closely. Figure 4.5 shows $-G$ plotted as a function of ζ for the case in which $\delta = 0.04$ and $\tau = 3$. These values are representative for fresh cracking catalyst fluidized by nitrogen at a pressure of 10 bar and at room temperature. For other combinations of solids and gases we refer to Tables 4.1 and 4.2.

The function G describes the contribution of gas adsorption to the total potential energy. At a value of $\zeta = 1.1776$ G is zero and at still lower values of ζ it becomes negligibly small (see Fig. 4.5). This is easily explained since at these small distances there is not enough space for the gas molecules to creep into the fissure between the particles (or between a particle and a flat plane). For higher values of ζ, $(-G)$ becomes negative since the adsorption energy is released. Now $(-G)$ increases rapidly to reach a maximum at $\zeta = \sqrt{2} = 1.4142$. At still higher values of ζ, $(-G)$ decreases again since the contribution of gas adsorption to the total binding energy decreases to become zero at infinite distance.

We now return to the function χ_t. It is plotted in Fig. 4.6—for the same reason as given in Section 4.3—versus the actual dimensionless distance ω between the particles. For the additional relevant parameters the values $\alpha = 7$ and $\psi = 2$ were chosen. The calculation route is as follows.

Choose a value of ζ_1 and calculate the corresponding value of the function G as well as $dG/d\zeta$ and $\int_\zeta^\infty Gd\zeta_1$. Next derive the corresponding value of η from eqn (4.20) after which χ_t can be calculated from eqn (4.19). χ_t can be plotted against ω and the dimensionless force can be derived by differentiation with respect to ω.

A first glance at Fig. 4.6 will make it clear that the number of mathematical solutions of χ_t is larger than one. This is caused by the fact that ω itself is a complicated function of ζ, in some regions

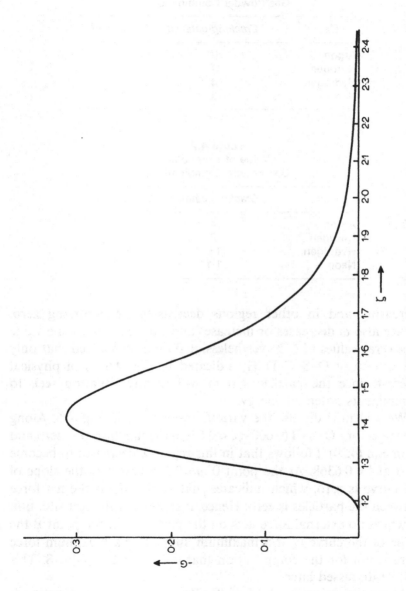

Fig. 4.5. The adsorption function *G* versus ζ.

Table 4.1
Value of α for some
Gas/Powder Combinations

Gas	Cracking catalyst	Quartz sand
Argon	10	9
Nitrogen	7	6
Hydrogen	4	3
Neon	3	3

Table 4.2
Value of τ for some
Gas/Powder Combinations

Gas	Cracking catalyst	Quartz sand
Argon	4	4
Nitrogen	3	3
Hydrogen	1·5	1·6
Neon	1·1	1·3

increasing and in other regions decreasing or becoming zero. Hence also ω decreases or increases and can reach the same value at several values of ζ. Nevertheless, it should be realized that only the trajectory O–S–C–D–H, indicated in Fig. 4.6, is of physical interest since the particle system will in any situation seek to minimize its potential energy.

We will now discuss the various trajectories in Fig. 4.6. Along the trajectory O–S–B $(0·600 < \zeta < 0·6368)$ the function G is zero and from eqn (4.20) it follows that in this region η decreases to become zero at $\zeta = 0·6368$. At the point 0 $\omega = 0·105$ and here the slope of the curve is zero, which indicates that at this point the net force between the particles is zero. Hence, it represents the rest situation in which no external force acts on the particles. At the point B the slope of the curve is at a maximum, indicating a maximum force were it not for the complication that occurs at the point S. This will be discussed later.

Along the trajectory B–M–K $(0·6368 < \zeta < 1·1776)$ the function G is still zero while also η is zero everywhere. Hence the whole

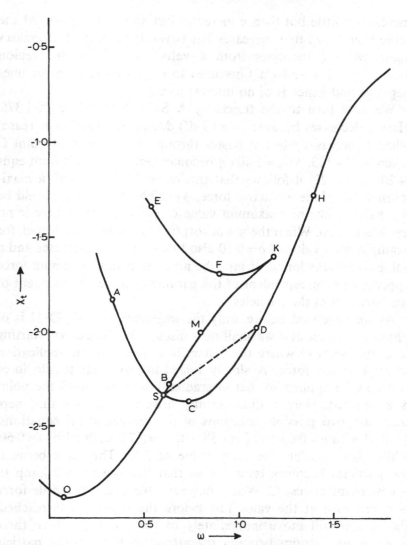

Fig. 4.6. The dimensionless interaction potential χ_t versus the distance parameter ω for $\psi = 2$.

trajectory O–S–B–M–K is, except for a constant difference $-\int_0^\infty G \cdot d\zeta_1$, the same as in the case of no gas adsorption.

At the point K ($\zeta = 1\cdot1776$) the effect of gas adsorption begins to play a role so that η becomes $\neq 0$, at first slowly so that ω still increases a little but then η increases fast and ω decreases. At the same time $(-\chi_t)$ first increases but beyond the point F decreases again when ζ increases from a value of 1·22. In the region $1·24 < \zeta < 1·37$, $(-\alpha dG/d\zeta)$ becomes so large that ω even becomes negative and hence is of no interest to us.

We now turn to the trajectory A–S–C–D–H where $\zeta > 1·376$. Here η decreases because $(-\alpha dG/d\zeta)$ decreases, hence ω increases when ζ increases while χ_t passes through a minimum at point C where $\omega = 1·383$. At $\zeta = 1·409$ η becomes zero again and from eqns (4.20) and (4.22) it follows that this coincides again with a maximum value of the attractive force, $\phi = 2·706$. This value should be compared with the maximum value of $\phi = 1·85$ when there is no gas adsorption. When the gas adsorption is stronger, indicated, for example, by a value of $\delta = 0·10$ also this maximum increases and a value of 3·804 is found. It must be noted that this maximum force appears to be independent of the parameter ψ, i.e. independent of the hardness of the particles.

As we have said before, only the trajectory O–S–C–D–H is of physical interest and we shall now discuss this trajectory, starting from the point O where the system is at rest without application of an external force. A slowly increasing external tensile force will now be applied so that ω gradually increases until the point S is reached. Here a discontinuity arises; it appears that here there are two possible solutions of the mathematical equations: $\zeta = 0·63$ with $\eta = 0·03$ and $\zeta = 1·38$ with $\eta = 0·78$, both with $\omega = 0·60$, while also $-\chi_t$ has the same value of 2·33. The force between the particles becomes repulsive so that the system will jump to a new point of rest C. When, however, the external tensile force is maintained at the value just before the point S was reached, the system will move immediately to the point D where there is again equilibrium between the attractive force of the particle system and the external force (the same slope). What actually happens at the point S is that the fissure between the particles suddenly widens while the particles are further flattened so that gas molecules can creep into this fissure. Although the potential energy is the same in both configurations, the latter is more favourable as it gives the possibility to reach a configuration with a lower potential energy at the point C.

At further increase of the tensile force ω will further increase until the point H is reached where the force reaches a maximum. Still further increase of the tensile force results in definite separation of the two surfaces and hence the cohesion is determined by the critical tensile force at the point H. When on the other hand the tensile force is reduced slowly from the point H, the system will move along the trajectory H–D–C to reach equilibrium at the point C. Hence, it is clear that the process of increasing and subsequently decreasing the tensile stress shows the phenomenon of hysteresis.

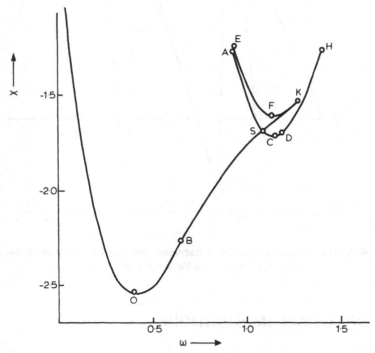

Fig. 4.7. The dimensionless interaction potential χ_t versus the distance parameter ω for $\psi = 10$.

In Fig. 4.7 χ_t is plotted as a function of ω in the case $\psi = 10$. The points and trajectories which correspond with those in Fig. 4.6 are indicated.

In Fig. 4.8 the dimensionless force ϕ is given as a function of ω for both $\psi = 2$ and $\psi = 10$, all with $\alpha = 7$, $\tau = 3$ and $\delta = 0.04$.

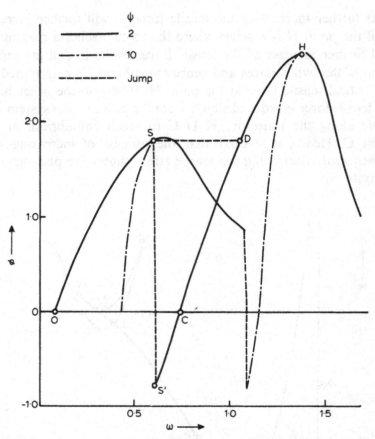

Fig. 4.8. The dimensionless force between two particles ϕ versus the distance parameter ω for $\psi = 2$ and $\psi = 10$.

4.5 PRACTICAL IMPLICATIONS

4.5.1 Powder Cohesion

The cohesion σ_c of powders is defined as the maximum tensile stress that a powder can withstand. It is directly related to the maximum cohesion force F_c at the contact point between two neighbouring particles.

According to Rumpf (1958) and Molerus (1975) this relation between σ_c and F_c is given by

$$\sigma_c = \frac{(1-\varepsilon)kF_c}{\pi d_p^2} \tag{4.23}$$

In this relation k is the coordination number defined as the number of contact points a particle has with other particles surrounding it. It decreases strongly at increasing porosity and together with the factor $(1-\varepsilon)$ it determines the porosity dependence of the tensile stress at F_c is independent of the porosity. If gas adsorption can be neglected (at low gas pressure) it is derived in sections 4.2 and 4.3 that

$$F_c = 1.85 \frac{AD}{12r_{ss}^2}$$

For most particle contacts it may be assumed that d_1 is equal to the diameter d_a of an asperity and d_2 equal to the particle diameter d_p. As generally $d_p \gg d_a$ it follows that $D = d_a$.

When we consider similar particles that only differ in average particle size the average diameter of the surface asperities can be expected to be equal. Hence, F_c will be the same for these powders and the tensile stress σ_c will be inversely proportional to d_p^2.

For a typical powder such as cracking catalyst it is found that $A = 16 \times 10^{-20}$ J (see Visser (1972)) while r_{ss} may be assumed to be 3.5×10^{-10} m. Taking $D = d_a = 5 \times 10^{-8}$ m it is found that $F_c = 10^{-8}$ N.

With an average particle diameter $d_p = 5 \times 10^{-5}$ m and a porosity of 0.40 while $k = 8$ it follows that $\sigma_c = 6.1$ N/m². This is a reasonable value for cracking catalyst at room temperature and a gas pressure of 1 bar.

The effect of gas adsorption on the cohesion between particles and hence on the tensile strength can be proved by increasing the gas pressure, since this causes an increase of the gas adsorption. This was experimentally demonstrated at Eindhoven by Piepers et al. (1984) with a specially designed tilting fluidized bed that could be operated at gas pressures up to 10 bar. In this apparatus the cohesion constant of a powder can be determined by measuring the critical tilting angle just before shearing off of the bed at various normal stresses at the bottom of the bed. By extrapolating to a normal stress of zero the cohesion constant is found.

When this is done for a not very cohesive powder it is necessary to operate the bed at rather low bed heights of, for example, 0·02 m. In order to further reduce the normal stress at the bottom, one can apply an upward gas flow at a velocity below the incipient fluidization velocity. Thus, the critical tilting angle can be determined at various gas velocities and hence at various normal pressures. This was done with cracking catalyst as a powder and argon as the fluidization gas at various gas pressures. The results are presented in Fig. 4.9. From these curves it follows clearly that the cohesion constant (i.e. the shear stress at $\sigma_\eta = 0$) increases with the gas pressure. At a pressure of 1 bar the cohesion constant appears to be about $4 \, \mathrm{N \, m^{-2}}$ which is in fair agreement with the above-derived value for the tensile strength ($6.1 \, \mathrm{N \, m^{-2}}$) assuming that at this gas pressure gas adsorption can be neglected.

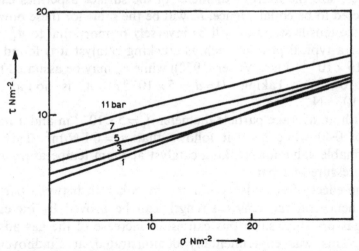

Fig. 4.9. Experimental yield curves of fresh cracking catalyst at various gas pressures of argon.

In dimensionless notation the theoretical cohesion ϕ_{max} as derived from the theory presented in this chapter, at various gas pressures of argon, is correlated with the measured cohesion constant as mentioned above (see Fig. 4.10). A remarkably good correlation between the two quantities can be observed.

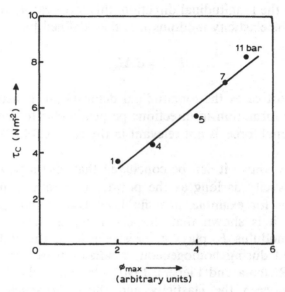

Fig. 4.10. The theoretical cohesion ϕ_{max} versus the experimental cohesion constant τ_c (parameter is pressure in bar).

4.5.2 Powder Elasticity

In Chapter 1 it was argued that in most powder operations the powder retains a mechanical structure which, although weak, is of great importance for understanding the behaviour of the powder. Now all mechanical structures, whether weak or strong, have an intrinsic property generally called elasticity. This indicates that when an external force is applied the material can be deformed a little without losing its internal cohesion. At the same time the material generates an elastic force which neutralizes the external force. When the external force is removed the material returns to its original structure as long as the external force is not strong enough for permanent plastic deformation to result. The larger the external force the larger the deformation; the proportionality constant is called the elasticity modulus. When we consider a small rectangular block of the elastic material with a cross-section A and a length L on which an external force F is working in the longitudinal direction, this force causes a deformation ΔL. The elasticity modulus E is then defined by

$$F/A = E \cdot \Delta L/L$$

That in most cases the longitudinal deformation is accompanied by some deformation in directions perpendicular to the direction of the external force, is not relevant in the context of the following discussion.

From the above it can be concluded that also a powder has a certain elasticity as long as the powder maintains a mechanical structure as for example, in a fluidized bed of that powder. In Chapter 7 it is shown that this elasticity, as measured by the elasticity modulus E, plays a decisive role in the stability of a powder bed during homogeneous fluidization (see also a recent paper by Rietema and Piepers, 1990). Obviously there must be a relation between the elasticity and the interparticle forces as derived in this chapter. From Fig. 4.3 it will be clear that as long as two neighbouring particles are not moved too far apart the attractive force between the particles increases with their distance and hence the particle contact has elastic properties. The corresponding elasticity modulus can be defined as

$$E_{pc} = F_c/d_p \cdot \Delta L$$

in which F_c is the maximum cohesion force between the particles and $\Delta L = (\Delta \omega) \cdot r_{ss}$, which is the maximum distance over which the particles can be moved apart without losing contact. It will be clear also from Fig. 4.3 that the harder the particles the smaller ΔL will be. Hence, the particle contact elasticity E_{pc} depends on the hardness of the particles although the cohesion force is independent of the hardness.

When evaluating E_{pc} from the above equation with the data mentioned before $(F_c = 10^{-8}\,\text{N}, \quad L = 2 \times 10^{-10}\,\text{m}$ and $d_p = 50 \times 10^{-6}\,\text{m})$, one finds $E_{pc} = 10^6\,\text{N m}^{-2}$. This seems extremely high compared to data on the powder elasticity modulus E as found in Chapter 7, where E varies from about $10\,\text{N m}^{-2}$ to $0.1\,\text{N m}^{-2}$.

A similar problem, however, appears in the mechanics of a spiral spring which has an effective elasticity modulus E_s related to the original material elasticity modulus E_m by the equation

$$E_s/E_m = \frac{1-e}{128\pi}\left(\frac{d}{R}\right)^4$$

in which

e = free space between the separate windings,
d = diameter of the spring wire, and
R = radius of the spiral spring

(see *Dubbels Taschenbuch*, 1953). With $e = 0.5$ and $d/R = 0.2$ it follows that $E_s/E_m = 8 \times 10^{-6}$.

In Chapter 8 a powder model will be presented from which follows the real relation between E and E_{pc} and which predicts a relation between the two quantities of the right order of magnitude. The dependence of E on the porosity also follows from this model.

REFERENCES

Boehme, G., Krupp, H., Rabenhorst, H. & Sandstede, G., (1962). Adhesion measurements involving small particles. *Trans. Inst. Chem. Engrs*, **40**, 252.

Bradley, R.S. (1932). On the cohesive forces between solid surfaces and the surface energy of solids. *Phil. Mag.*, **13**, 853.

Cottaar, E.J.E. & Rietema, K. (1986). A theoretical study of the influence of gas adsorption on interparticle forces in powders. *J. Colloid Interface Sci.*, **109**, 249.

Dahneke, B. (1972). The influence of flattening on the adhesion of particles. *J. Colloid Interface Sci.*, **40**, 1.

Derjaguin, B.V., Muller, V.M. & Toporov, Y.P. (1975). Effect of contact deformations on the adhesion of particles. *J. Colloid Interface Sci.*, **53**, 314.

Donsi, G. & Massimilla, L. (1973). Particle to particle forces in fluidization of fine powders. *Proc. Int. Symp. Fluidization and its Applications*, Toulouse, France, p. 41.

Dubbels Taschenbuch (1953). Vol. 1, eleventh edition, p. 396.

Hamaker, H.C. (1937). The London–Van der Waals attraction between spherical particles. *Physica*, **4**, 1058.

Hertz, H. (1895). *Gesammelte Werke*, Leipzig, Germany.

Johnson, K.L., Kendall, K. & Roberts, A.D. (1971). Surface energy and the contact of elastic solids. *Proc. Roy. Soc. Lond.*, **A324**, 301.

Krupp, H. (1967). Particle adhesion, theory and experiment. *Advan. Colloid Interface Sci.*, **1**, 111.

Lennard-Jones, J.E. (1937). The equation of state of gases and critical phenomena. *Physica*, **4**, 941.

London, F. (1937). The general theory of molecular forces. *Trans. Faraday Soc.*, **33**, 8.

Massimilla, L. & Donsi, G. (1976). Cohesive forces in fluidization of fine particles. *Powder Techn.*, **15**, 253.

Molerus, O. (1975). Theory of yield of cohesive powders. *Powder Techn.*, **12**, 259,

Piepers, H.W., Cottaar, E.J.E., Verkooyen, A.H.A. & Rietema, K. (1984). Effects of pressure and type of gas on particle–particle interaction and the consequences for gas–solid fluidization behaviour. *Powder Techn.*, **37**, 55.

Pollock, H.M. (1978). Contact adhesion between solids in vacuum. II, Deformation and interfacial energy. *J. Phys. D, Appl. Phys.*, **11**, 39.

Rietema, K. & Piepers, H.W. (1990). The effect of interparticle forces on the stability of gas-fluidized beds. Part I, Experimental evidence. *Chem. Engr. Sci.*, **45**, 1627.

Rumpf, H. (1958). Grundlagen und Methoden des Granulierens. *Chemie-Ing. Techn.*, **30**, 144.

Tabor, D. (1977). Surface forces and surface interactions. *J. Colloid Interface Sci.*, **58**, 2.

Visser, J. (1972). On Hamaker constants. A comparison between Hamaker constants and Lifshitz–Van der Waals constants. *Advan. Colloid Interface Sci.*, **3**, 331.

4.A APPENDIX

4.A.1 The Interaction Energy between a Gas and a Solid

For the interaction energy between an individual gas molecule and an individual molecule of the solid we assume a Lennard–Jones potential:

$$V_{gs}(r) = C_{gs}\left(-\frac{1}{r^6} + \frac{r_{gs}^6}{2r^{12}}\right) \qquad (4.A.1)$$

The energy of a single gas molecule which is at a distance d from a solid plane is then given by

$$U_a(d) = 2\pi n_s \int_0^{\pi/2} \left(\int_{d/\cos\theta}^{\infty} V_{gs} r^2 \, dr\right) \sin\theta \, d\theta$$

$$= 2\pi n_s C_{gs}\left(-\frac{1}{6d^3} + \frac{r_{gs}^6}{144d^9}\right) \qquad (4.A.2)$$

This energy is at a minimum for $d = r_{gs}/\sqrt{2}$. The binding energy of a single gas molecule is therefore given by

$$U_a = -\frac{4\sqrt{2}\pi n_s C_{gs}}{9 r_{gs}^3} \qquad (4.A.3)$$

When the case of a single gas molecule between two solid planes at a distance y is considered, the energy is

$$U_c(y, d) = U_a(d) + U_a(y - d) \qquad (4.A.4)$$

For low values of y ($y < 2r_{gs}$) this function has a minimum at $d = \frac{1}{2}y$. For higher values two minima will occur. As long as y is not too large, the value of the energy in these minima does not differ very much from the value of $d = \frac{1}{2}y$. Therefore, we will write the binding energy between two planes as

$$U_c(y) = U_a(\tfrac{1}{2}y) + U_a(\tfrac{1}{2}y) = -\frac{4U_a}{\sqrt{2}}\left(-\frac{3r_{gs}^3}{y^3} + \frac{8r_{gs}^9}{y^9}\right) \qquad (4.A.5)$$

4.A.2 The Density of Adsorbed Gas Molecules

The system of the free gas and the adsorbed gas at a free surface is considered. It is assumed there is equilibrium. Then the chemical potential μ_g for the free gas must be equal to the chemical potential μ_a of the adsorbed gas:

$$\mu_g = \mu_a = U_a + kT \log \frac{N_a}{N_{oa} - N_a} \qquad (4.A.6)$$

where N_a is the number of adsorbed gas molecules and N_{oa} the number of adsorption sites at the free surface. The same holds, of course, for the adsorption between two planes at distance y. Thus

$$\mu_g = \mu_c = U_c(y) + kT \log \frac{N_c(y)}{N_{oc} - N_c(y)} \qquad (4.A.7)$$

If all considitions, e.g. temperature and pressure, are the same it therefore follows that

$$\frac{N_c(y)}{N_{oc} - N_c(y)} = \frac{N_a}{N_{oa} - N_a} \exp\{[-U_c(y) + U_a]/kT\} \qquad (4.A.8)$$

Since the surface covered by one gas molecule is of the order of r_{gg}^2 it follows that

$$N_c(y) = \frac{1}{r_{gg}^2} \frac{\delta \exp\{[-U_c(y) + U_a]/kT\}}{1 + \delta \exp\{[-U_c(y) + U_a]/kT\}} \qquad (4.A.9)$$

using the definition

$$\delta = \frac{N_a}{N_{oa} - N_a} \qquad (4.A.10)$$

Since the chemical potential of a free, ideal gas is proportional to the logarithm of the gas pressure, it follows from eqn (4.A.6) that δ also is proportional to the pressure.

5

Two-phase Systems. The Equations of Motion

NOTATION

A	Cross-sectional area (m^2)
B	Buoyancy (N m^{-2})
C	Correlation coefficient (—)
d_p	Particle size (m)
$f(\varepsilon)$	Function of the porosity (—)
F_c	Flux in the gas phase
F_d	Flux in the dispersed phase
$\mathbf{F}_{dc}$	Force per unit volume dispersed phase (N m^{-3})
$\mathbf{F}_{pc}$	Force on a single particle (N)
$\mathbf{F}_s$	Slip force (N m^{-2})
g	Gravitational acceleration (m s^{-2})
$h(\varepsilon)$	Function of the porosity (—)
m_i	Mobility of solid i (m^4(N s)$^{-1}$)
n	Normality vector (—)
N	Number of particles in control volume (m^{-3})
N_g	Gas–solid interaction number (—)
p	Gas pressure (N m^{-2})
r_0	Size of control volume (m)
$\mathbf{R}_c$	Turbulent momentum transport tensor in the gas phase (N m^{-2})

95

R_d Turbulent momentum transport tensor in the dispersed
          phase (N m^{-2})
S Surface area of particle (m^2)
t Time (s)
v_c Velocity vector in the gas phase (m s^{-1})
v_d Velocity vector in the dispersed phase (m s^{-1})
v_s Slip velocity (m s^{-1})
V Volume of dispersed system (m^3)
V_a Characteristic velocity of handling apparatus (m s^{-1})
V_p Volume of a particle (m^3)
x, y, z Rectangular coordinates (m)

α_i Volume fraction of solid i (—)
δ Unit tensor (—)
ε Porosity (—)
μ Gas viscosity (N s m^{-2})
ρ_c Density of gas phase (kg m^{-3})
ρ_d Density of dispersed phase (kg m^{-3})
ρ_m Density of surrounding medium (kg m^{-3})
τ_c Stress tensor in the gas phase (N m^{-2})
σ Stress tensor in the dispersed phase (N m^{-2})

5.1 INTRODUCTION

A powder is a dense population of fine solid particles, the interstitial space of which is filled with a gas that can move more
or less freely between the particles. A powder, therefore, should
be considered as a two-phase system of solid particles and a gas
in which both phases can move separately at different velocities.

The velocity field of the solid particles is related to the powder
stress tensor σ already discussed in Chapter 3 for the case of
a powder at rest. Similarly, the velocity field of the gas phase
is related to the fluid shear tress tensor τ_c pertaining to this
phase. It must be noted that the tensor τ_c is not quite comparable with the tensor σ as it is normal practice to bring the isotropic gas pressure p outside the tensor. Since it is generally

agreed that the powder stress, unlike the fluid shear stress, has no direct effect upon the other phase, this is usually not done for the powder stress. The two velocity fields are coupled by the interaction force between the two phases. This interaction force is due to the action of the shear tensor τ_c and the gas pressure p upon the powder particles, while the velocity field between the two phases causes an extra friction upon the particles. Only when the solid particles are relatively large and/or the gas viscosity is low can the interaction between the two phases be neglected and the concept of the powder as a two-phase system becomes of no importance. It will be shown in Chapter 11 that this is roughly the case when the dimensionless gas–solid interaction number

$$N_g = \frac{\rho_d d_p^2 g}{\mu V_a} > 100$$

Here V_a is a characteristic velocity during the process of powder handling or of the apparatus in which the powder is treated.

For a full understanding of a powder-treating process it is necessary to start from the equations of motion of the system. These comprise two mass balances or continuity equations, one for each phase; and two momentum equations, again one for each phase. In these four differential equations four variables occur, viz. the porosity, the pressure, and the two velocities of the participating phases. Hence, when the boundary conditions are known the problem of the motion of the powder is defined and can in principle be solved, which means that it can be established how these quantities vary with time and space.

5.2 AVERAGING OF VARIABLES

During the handling of powders momentary point variables such as the two velocities, the pressure in the gas phase and the porosity vary owing to macroscopic variations that determine the overall macroscopic behaviour of the powder at the scale of the macroscopic equipment in which it is contained.

On top of these macroscopic variations the variables also fluctuate in time and space owing to microscopic variations at the scale of the powder particles themselves. These variations arise from the complicated paths of the individual particles and from the distortion of fluid streamlines around and between the solid particles. These variations are strongly stochastic (microturbulence) but when suitably averaged will turn out to be zero. Because of this stochastic character it is impossible—but also meaningless—to set up a mathematical model on the basis of which these fluctuations can be calculated.

As we are mainly interested in the overall behaviour of the powder, averaging of the variables is necessary. We therefore consider any momentary point variable a' as the superposition of its average value a and its fluctuation a'' ($a' = a + a''$). When the averaging is properly done the average of a'' should be zero.

We propose to take the average over a cubic control volume element $\Delta V = r_0^3 = \Delta x \cdot \Delta y \cdot \Delta z$ in case we are dealing with a scalar variable or a body force, and over a square control surface area $\Delta A = r_0^2$ when we are dealing with a vector variable or a flux.

The control volume ΔV is partly occupied by gas ($= \Delta V_c$) and partly by dispersed solid particles ($= \Delta V_d$), so $\Delta V = \Delta V_c + \Delta V_d$. We can now introduce the porosity as

$$\varepsilon = \frac{1}{\Delta V} \int_{\Delta V_c} \mathrm{d}x \cdot \mathrm{d}y \cdot \mathrm{d}z$$

where the integration is extended only over the part of V that is occupied by the gas phase. The average of a concentration a'_c in the gas phase is now defined by

$$\varepsilon a_c = \frac{1}{\Delta V} \int_{\Delta V_c} a'_c \cdot \mathrm{d}x \cdot \mathrm{d}y \cdot \mathrm{d}z$$

and similarly a concentration a'_d in the solid phase by

$$(1 - \varepsilon)a_d = \frac{1}{\Delta V} \int_{\Delta V_c} a'_d \cdot \mathrm{d}x \cdot \mathrm{d}y \cdot \mathrm{d}z$$

When averaging fluxes, however, these definitions are not adequate. In that case we must reconsider the definition of the porosity.

Suppose we want to average a flux F'_{cx} which pertains to the gas phase. F'_{cx} is a flux in the x-direction and is the x-component of a flux F'_c. We now put

$$\varepsilon^{(x)} = \frac{1}{\Delta y \cdot \Delta z} \int_{\Delta A_c} dy \cdot dz$$

but we shall show that $\varepsilon^{(x)} = \varepsilon$. To this end we suppose that the control volume ΔV is split up into a large number of thin slices perpendicular to the direction of the flux F_{cx} where the thickness of these slices is small compared to the size of the powder particles. The intersection of these slices with the particles then can be conceived as being constant (independent of x, see Fig. 5.1). It

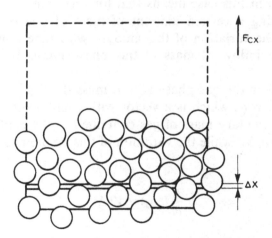

Fig. 5.1. Illustration of the proof that the volume porosity is equal to the cross-section porosity.

follows that ε does not change if we put

$$\varepsilon = \frac{1}{\Delta x} \int_{\Delta x} \left(\frac{1}{\Delta y \cdot \Delta z} \int_{\Delta A_c} dy \cdot dz \right) dx = \frac{1}{\Delta y \cdot \Delta z} \int_{\Delta A_c} dy \cdot dz = \varepsilon^{(x)}$$

In accordance with the foregoing we now put

$$\varepsilon F_{cx} = \frac{1}{\Delta A} \int_{\Delta A_c} F'_{cx} \cdot dy \cdot dz$$

Similarly for a flux of the dispersed solid phase

$$(1 - \varepsilon) F_{dx} = \frac{1}{\Delta A} \int_{\Delta A_a} F'_{dx} \cdot dy \cdot dz$$

Also for the y- and z-components of F_c and F_d similar definitions hold.

5.3 THE MASS BALANCES

The mass balances are based on the principle of conservation of mass, which in this case means that for a certain control volume ΔV containing a certain amount of mass of one of the phases concerned the variation of this amount with time must be equal to the total inflow of mass of this phase minus the total outflow.

We consider the gas phase with a mass density ρ_c and a local linear velocity v'_c, which is a vector with components v'_{cx}, v'_{cy} and v'_{cz}. The momentary mass of gas contained in the control volume is equal to $\varepsilon \rho_c \Delta V$ while the variation of this amount in the time Δt is

$$\frac{\partial}{\partial t}(\varepsilon \rho_c) \cdot \Delta V \cdot \Delta t$$

The total inflow of gas during time Δt is equal to

$$\left\{ \int_{(\Delta y, \Delta z)_c x = x_1} \rho_c v'_{cx} \cdot dy \cdot dz + \int_{(\Delta x, \Delta z)_c y = y_1} \rho_c v'_{cy} \cdot dx \cdot dz \right.$$

$$\left. + \int_{(\Delta x, \Delta y)_c z = z_1} \rho_c v'_{cz} \cdot dx \cdot dy \right\} \Delta t$$

while the total outflow in this time is

$$\left\{ \int_{(\Delta y, \Delta z)_c x = x_1 + \Delta x} \rho_c v'_{cx} \cdot dy \cdot dz + \int_{(\Delta x, \Delta z)_c y = y_1 + \Delta y} \rho_c v'_{cy} \cdot dx \cdot dz \right.$$

$$\left. + \int_{(\Delta x, \Delta y)_c z = z_1 + \Delta z} \rho_c v'_{cz} \cdot dx \cdot dy \right\} \Delta t$$

We assume that the density ρ_c does not vary over ΔV, and hence can be brought outside the integrals; further, we apply the definitions of averages discussed in Section 5.2 and the mathematical definition that for a variable b

$$\lim_{\Delta x \to 0} (b_{x+\Delta x} - b_x) = \frac{\partial b}{\partial x} \Delta x$$

After dividing by $\rho_c \Delta x \cdot \Delta y \cdot \Delta z \cdot \Delta t$ we can now write the mass balance as

$$\frac{\partial}{\partial t}(\varepsilon) = -\left\{ \frac{\partial}{\partial x}(\varepsilon v_{cx}) + \frac{\partial}{\partial y}(\varepsilon v_{cy}) + \frac{\partial}{\partial z}(\varepsilon v_{cz}) \right\}$$

The expression between braces is generally simply written as $\text{div}(\varepsilon v_c)$ or as $\nabla \cdot (\varepsilon v_c)$ in which ∇ is the so-called nabla operator which has a vector character:

$$\nabla = \left(\frac{\partial}{\partial x}, \frac{\partial}{\partial y}, \frac{\partial}{\partial z} \right)$$

$\nabla \cdot (\varepsilon v_c)$ indicates the scalar product of the vector ∇ and the vector (εv_c).

The mass balance for the gas phase now reads simply:

$$\frac{\partial}{\partial t}(\varepsilon) + \nabla \cdot (\varepsilon v_c) = 0 \tag{5.1}$$

Similarly the mass balance for the dispersed solid phase reads:

$$\frac{\partial}{\partial t}(1-\varepsilon) + \nabla \cdot \{(1-\varepsilon)v_d\} = 0 \tag{5.2}$$

Remark: When the densities are not constant during the time Δt these densities must be left under the integral and differential operators. In that case averaging can be applied as well, but the averages thus defined need not be the same as those defined in Section 5.2 as generally

$$\int_{\Delta y, \Delta z} \rho_c v'_{cx} \, \mathrm{d}y \, \mathrm{d}z \neq \rho_c \int_{\Delta y, \Delta z} v'_{cx} \, \mathrm{d}y \, \mathrm{d}z$$

When, however, we reduce the timespan Δt the variation of the density will become so small that the definitions of Section 5.2 can safely be used.

5.4 THE MOMENTUM BALANCES

The momentum balances are also based on a law of conservation, viz. the conservation of momentum. A specific volume under consideration contains at a certain moment a certain amount of momentum which is equal to the total mass contained in this control volume times the momentary velocity of that mass. Hence the concentration of that momentum is equal to the mass density times the momentary local velocity.

This also holds for a two-phase system on the understanding that the total amount of momentum can be split into the amount of momentum of one of the two phases and that of the other phase. This means that for each of the two phases a momentum balance should be drawn up.

Of each phase the momentum concentration in a control volume can change by:

(a) flow of mass of that phase in or out of the control volume, since these mass flows also imply a flux of momentum;
(b) external forces acting upon this phase, such as
  (1) viscous shearing stresses,
  (2) pressure gradients,
  (3) body forces, the most prominent of which is the gravity

force. In certain specific cases also centrifugal forces, electrostatic and magnetic forces fall in this category;

(c) internal forces, viz. the viscous shearing stress which the two phases exert upon each other when they have different velocities.

The momentum balance of each phase now reads in principle:

(rate of accumulation of momentum of phase considered)
$$= \text{(rate-in of momentum by convection)}$$
$$- \text{(rate-out of momentum by convection)} \qquad (5.3)$$
$$+ \text{(sum of external forces acting upon that phase)}$$
$$+ \text{(sum of internal forces acting upon that phase)}$$

It will be clear that the internal force is the same for each phase but for the sign. Hence, when $\mathbf{F}$ is the force which phase 1 exerts upon phase 2, then $-\mathbf{F}$ is the force which phase 2 exerts upon phase 1.

5.4.1 Momentum Balance of the Solid Phase

As $\mathbf{v_d}$ is the local momentary concentration of momentum of the solid phase with components $\rho_d v'_{dx}$, $\rho_d v'_{dy}$ and $\rho_d v'_{dz}$ the average concentration of momentum in the x-direction equals

$$\int_{\Delta Vd} \rho_d v'_{dx} \cdot dx \cdot dy \cdot dz = (1-\varepsilon)\rho_d v_{dx}$$

and the rate of accumulation of momentum in this direction equals

$$\frac{\partial}{\partial t}\{(1-\varepsilon)\rho_d v_{dx}\}$$

Similar expressions are found for the accumulation of the y- and z-components of momentum. Hence, the total rate of accumulation is given by

$$\frac{\partial}{\partial t}\{(1-\varepsilon)\rho_d \mathbf{v_d}\}$$

The local rate-in of momentum is equal to the local concentration of momentum times the local velocity, i.e. $(\rho_d v_d')v_d'$ which is a tensor with nine components:

$$\begin{pmatrix} (\rho_d v_{dx}')v_{dx}', & (\rho_d v_{dx}')v_{dy}', & (\rho_d v_{dx}')v_{dz}' \\ (\rho_d v_{dy}')v_{dx}', & (\rho_d v_{dy}')v_{dy}', & (\rho_d v_{dy}')v_{dz}' \\ (\rho_d v_{dz}')v_{dx}', & (\rho_d v_{dz}')v_{dy}', & (\rho_d v_{dz}')v_{dz}' \end{pmatrix}$$

The total rate-in of momentum in the x-direction is found after integration over the faces of the control volume:

$$\int_{(\Delta y, \Delta z)_d x = x_1} (\rho_d v_{dx}')v_{dx}' \cdot dy \cdot dz + \int_{(\Delta x, \Delta z)_d y = y_1} (\rho_d v_{dx}')v_{dy}' \cdot dx \cdot dz$$

$$+ \int_{(\Delta x, \Delta y)_d z = z_1} (\rho_d v_{dx}')v_{dz}' \cdot dx \cdot dy = [(1-\varepsilon)\rho_d v_{dx}v_{dx}]_{x=x_1}\Delta y \cdot \Delta z$$

$$+ [(1-\varepsilon)\rho_d v_{dx}v_{dy}]_{y=y_1}\Delta x \cdot \Delta z + [(1-\varepsilon)\rho_d v_{dx}v_{dz}]_{z=z_1} \cdot \Delta x \cdot \Delta y$$

$$+ \int_{(\Delta y, \Delta z)_d x = x_1} (\rho_d v_{dx}'')v_{dx}'' \cdot dy \cdot dz + \int_{(\Delta x, \Delta z)_d y = y_1} (\rho_d v_{dx}'')v_{dy}'' \cdot dx \cdot dz$$

$$+ \int_{(\Delta x, \Delta y)_d z = z_1} (\rho_d v_{dx}'')v_{dz}'' \cdot dx \cdot dy$$

Averaging a flux b' was defined so that the average of the fluctuation b'' became zero. This, however, means that the average of a product of fluctuations, such as $(\rho_d v_{di}'')v_{dj}''$ most probably will not be zero, hence the last three terms of the above expression cannot be neglected. In order to account for this effect we define a tensor

$$R_d = \overline{(1-\varepsilon)\rho_d v_d'' v_d''}$$

so that the last three terms can be put equal to:

$$[R_d]_{x,x_1} \cdot \Delta y \cdot \Delta z, \quad [R_d]_{x,y_1} \cdot \Delta x \cdot \Delta z \quad \text{and} \quad [R_d]_{x,z_1} \cdot \Delta x \cdot \Delta y$$

For the total rate-out of momentum in the x-direction a similar expression is found as for the total rate-in, except that the evaluation of all six terms is performed now at the levels $x_1 + \Delta x$, $y_1 + \Delta y$ and $z_1 + \Delta z$. Combination of rate-in and rate-out now gives the net rate of momentum in the x-direction by convection:

$$-(\rho_d \{\nabla \cdot [(1 - \varepsilon)\mathbf{v}_d \mathbf{v}_d]\}_x + [\nabla \cdot \mathbf{R}_d]_x)\Delta x \cdot \Delta y \cdot \Delta z$$

Similar expressions are found for the net rates of momentum in the y- and z-directions.

The tensor $\mathbf{R}_d$ needs some further explanation. It accounts for some degree of coupling between neighbouring particle streampaths: momentum of particles along one path is exchanged with momentum of particles along another, neighbouring path owing to the tortuosity of these streampaths. This phenomenon also occurs in turbulent single-phase flow and the resulting stress is then called the 'Reynolds stress'. Its effect is similar to that of viscous momentum exchange and causes a flattening of velocity profiles. It can be accounted for by means of a formal increase of the viscosity.

In two-phase flow similar phenomena occur, in principle in both phases. They play a role especially at high porosities when the dispersed particles can move more or less freely in all directions. At lower porosities ($\varepsilon < 0.70$), however, the particles move mainly along parallel streampaths and in that case $\mathbf{R}_d$ can be neglected. The continuous phase on the other hand moves along streampaths which run around the dispersed particles, alternating through narrow and wide pores between the particles. These streampaths, hence, are tortuous to a rather high degree, which now causes a tensor $\mathbf{R}_c$ to arise in the analysis (comparable to $\mathbf{R}_d$, but now based on the fluctuations of the continuous phase velocity). This will be discussed further in Section 5.5 on the constitutive equations.

Before turning to the other contributions to the momentum accumulation, we must first discuss a problem connected with those particles that are lying partly inside and partly outside the cubic control volume and which hence are cut by the faces of this cubic volume element. Shear stresses acting on these particles will, hence, belong only partly to the total shear stresses on the particles inside the control volume. In order to deal with this problem we shall distinguish three groups of particles (see Fig. 5.2):

(a) a group N_0 of those particles that are lying completely inside the cubic volume element;

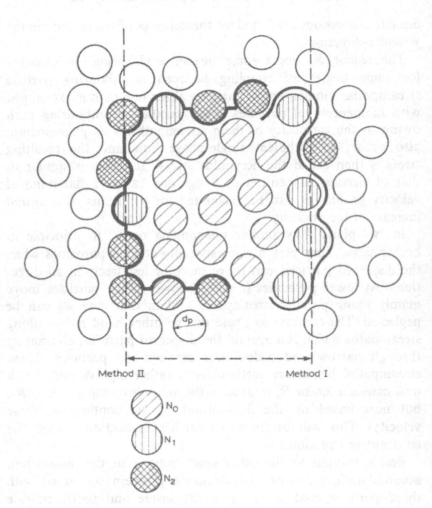

Fig. 5.2. Interface of control volume with indications of method I and method II.

(b) a group N_1 of particles that are cut by the faces of the cubic volume element but whose mass centres are lying inside the cubic volume element or in its faces;
(c) a group N_2 of particles cut by the faces but whose mass centres are outside the cubic volume element.

We shall now suppose that the boundaries of the control volume are flexible and run smoothly through the continuous phase around the particles N_1 without touching these particles while the particles N_2 stay outside the control volume (Fig. 5.2, method I).

Of course, another method of dealing with this problem is conceivable. In this method the control volume again contains the particles N_0 and N_1 but not the particles N_2. In this method, however, the boundaries of the control volume follow the exact mathematical planes ($x =$ constant, $y =$ constant, $z =$ constant) as long as they run through the continuous (gas) phase (see Fig. 5.2, method II, the thick line), but follow the inward part S_i of the surface of the particles N_2 facing the control volume and the outward-facing part S_e of the surface of the particles N_1. This method implies that the forces acting on particles that are intersected by the flat boundaries of the cubic volume element now belong partly to the external forces and partly to the internal forces.

In a paper by Rietema and Van den Akker (1983) on the momentum equations in dispersed two-phase systems this problem has been dealt with in a rather rigorous way. In that paper it is shown that both methods give exactly the same final result.

Method I, of course, can only be applied when the porosity is so high that the particles are indeed flowing freely and contacts between the particles occur only occasionally and can be neglected. Although this is not really possible with powders, the fact that both methods give the same final result allows us yet to apply here method I, which we prefer since with this method the derivation of the momentum equations is so much simpler.

It should be realized that whether we apply method I or method II does not affect the derivations—as given before—of the rate of momentum accumulation and of the net rate of momentum by convection.

For simplicity we shall assume that all particles have the same volume V_p and are uniformly distributed over the control volume. In that case it can be shown (see the paper by Rietema and Van den Akker) that

$$(N_0 + N_1)V_p = (1-\varepsilon)r_0^3 \qquad (5.4)$$

The internal force exerted by the dispersed phase upon the continuous phase, the so-called interaction force $\mathbf{F}_{dc}$ (per unit volume of the dispersion), is now found by summation of all forces that the individual particles contained in the control volume exert upon the continuous phase. Hence,

$$\mathbf{F}_{dc} = \sum_{\text{all particles}} \mathbf{F}'_{pc} \qquad (5.5)$$

In this equation $\mathbf{F}'_{pc}$ is the force on a single particle with an external surface area S_p:

$$\mathbf{F}'_{pc} = \int_{S_p} \mathbf{n} \cdot (p'\boldsymbol{\delta} + \boldsymbol{\tau}'_c)dS \qquad (5.6)$$

where $\mathbf{n}$ is the normal vector to the surface S_p of the particle and $\boldsymbol{\delta}$ is the unit tensor while $\boldsymbol{\tau}'_c$ and p' are respectively the local shear stress tensor and the local pressure in the continuous phase.

Applying Gauss's theorem it is found that

$$\mathbf{F}'_{pc} = \int_{V_p} (\nabla p' + \nabla \cdot \boldsymbol{\tau}'_c)\,dV_p \qquad (5.7)$$

The total interaction force in the control volume therefore is

$$r_0^3 \mathbf{F}_{dc} = \sum_{\text{all particles } N_0 + N_1} \int_{V_p} (\nabla p' + \nabla \cdot \boldsymbol{\tau}'_c)\,dV_p$$

and because of eqn (5.4) this can be written as

$$\mathbf{F}_{dc} = \frac{1}{r_0^3} \int_{(\Delta x, \Delta y, \Delta z)_d} (\nabla p' + \nabla \cdot \boldsymbol{\tau}'_c)\,dx\,dy\,dz \qquad (5.8)$$

The summation over all particles $N_0 + N_1$ means that the interaction force F'_{pc} is averaged over the dispersed phase in the control volume. It must be realized, however, that although the average of p'' is zero this does not mean that also the average of $(\nabla p'')$ is zero. The same holds true for the variation of τ'_c. Hence, when eqn (5.8) is elaborated there remains a term $\mathbf{F_s}$ where

$$\mathbf{F_s} = \frac{1}{r_0^3} \int_{(\Delta x, \Delta y, \Delta z)} (\nabla p'' + \nabla \cdot \tau''_c) \, dx \, dy \, dz \qquad (5.9)$$

Finally it is found that

$$\mathbf{F_{dc}} = (1 - \varepsilon)(\nabla p + \nabla \cdot \tau_c) + \mathbf{F_s} \qquad (5.10)$$

The force $\mathbf{F_s}$ can be ascribed to the variations of the pressure and the shear stresses around the particles when these particles have a velocity different from that of the continuous phase. Hence, this force can be expected to be related directly to the so-called slip velocity $\mathbf{v_s}$ which is defined as $(\mathbf{v_d} - \mathbf{v_c})$. $\mathbf{F_s}$ is generally called the slip force. It depends further on the viscosity of the continuous phase, on the particle size and on the porosity.

As in applying method I, the solid particles are entirely inside the control volume and there are no external forces acting upon the particles except the body force caused by gravitation. It will be clear that this latter force is $(1 - \varepsilon)\rho_d \mathbf{g} r_0^3$ and so we can now make up the total momentum balance for the dispersed phase:

$$\frac{\partial}{\partial t}\{(1 - \varepsilon)\rho_d \mathbf{v_d}\} = -\rho_d \quad \nabla \cdot \{(1 - \varepsilon)\mathbf{v_d}\mathbf{v_d}\} - \nabla \cdot R_d + (1 - \varepsilon)\rho_d \mathbf{g} - \mathbf{F_{dc}}$$

$$(5.11)$$

The term on the left-hand side and the first term on the right-hand side of this equation can—with the help of the continuity equation for the dispersed phase—be elaborated into

$$\rho_d(1 - \varepsilon)\left\{\frac{\partial}{\partial t}\mathbf{v_d} + \mathbf{v_d} \cdot \nabla \mathbf{v_d}\right\}$$

which is generally written as

$$\rho_d(1-\varepsilon)\frac{D}{Dt}\mathbf{v}_d$$

where the vector operator D/Dt stands for

$$\left(\frac{\partial}{\partial t}\cdots + \mathbf{v}_d\cdot\mathbf{\nabla}\cdots\right)$$

After substitution of $\mathbf{F}_{dc}$ the result is

$$\rho_d(1-\varepsilon)\frac{D}{Dt}\mathbf{v}_d = -\mathbf{\nabla}\cdot\mathbf{R}_d-(1-\varepsilon)\mathbf{\nabla}p-(1-\varepsilon)\mathbf{\nabla}\cdot\mathbf{\tau}_c+(1-\varepsilon)\rho_d\mathbf{g}-\mathbf{F}_s$$

$$(5.12)$$

Equation (5.12) was derived for the case of free floating solids (method I). This method does not allow for a derivation of the momentum equation for particles which are in permanent contact, as is the case when we are dealing with a powder. In that instance method I should be applied. As remarked before, however, method I and method II give the same result when the solids are free floating. When the particles are in permanent contact method II indicates that only a term $-\mathbf{\nabla}\cdot\mathbf{\sigma}$ should be added to the momentum balance. This term accounts for the friction, cohesion and stresses as they occur in the powder mass. The ultimate result is:

$$\rho_d(1-\varepsilon)\frac{D}{Dt}\mathbf{v}_d = -\mathbf{\nabla}\cdot\mathbf{\sigma}-(1-\varepsilon)\mathbf{\nabla}p-(1-\varepsilon)\mathbf{\nabla}\cdot\mathbf{\tau}_c+(1-\varepsilon)\rho_d\mathbf{g}-\mathbf{F}_s$$

$$(5.13)$$

5.4.2 Momentum Balance of the Continuous Phase

We shall now proceed with the derivation of the momentum balance of the continuous phase and derive the effects of the external force caused by the shear stresses in this phase. As we apply method I in which the boundaries of the control volume are lying anywhere in the continuous phase, it follows that in averaging the shear stresses over the faces of the control volume the porosity does not come in.

The fluid shear stress τ_c is analogous to the powder stress. Both have nine components, but in the case of the fluid shear stress the isotropic pressure p has been eliminated by deduction of the tensor $p\delta$ in which δ is the unit tensor. This means that $\sigma_{xx} + \sigma_{yy} + \sigma_{zz} = 0$. In the x-direction the stresses τ'_{xx}, τ'_{xy} and τ'_{xz} operate. Hence, the rate-in of momentum in the x-direction

$$\int_{(\Delta y, \Delta z)_c \, x = x_1} \tau'_{xx} \, dy \, dz + \int_{(\Delta x, \Delta z)_c \, y = y_1} \tau'_{xy} \, dx \, dz + \int_{(\Delta x, \Delta y)_c \, z = z_1} \tau'_{xz} \, dx \, dy$$

After averaging this becomes

$$\tau_{xx} \Delta y \Delta z \underset{\text{at } x = x_1}{} + \tau_{xy} \Delta x \Delta z \underset{\text{at } y = y_1}{} + \tau_{xz} \Delta x \Delta y \underset{\text{at } z = z_1}{}$$

Similarly the rate-out in this direction is

$$\tau_{xx} \Delta y \Delta z \underset{\text{at } x_1 + \Delta x}{} + \tau_{xy} \Delta x \Delta z \underset{\text{at } y_1 + \Delta y}{} + \tau_{xz} \Delta x \Delta y \underset{\text{at } z_1 + \Delta z}{}$$

Together these contributions give

$$-\left\{ \frac{\partial}{\partial x} \tau_{xx} + \frac{\partial}{\partial y} \tau_{xy} + \frac{\partial}{\partial z} \tau_{xz} \right\} \Delta x \cdot \Delta y \cdot \Delta z = -r_0^3 [\nabla \cdot \tau_c]_x$$

Analogous expressions are found for the contributions in the y- and z-directions.

The other external forces acting upon the continuous phase contained in the control volume include the action of the pressure p and the body force due to gravity.

The external force exerted in the x-direction by the pressure is acting only upon the faces $(\Delta y, \Delta z)$. When we again apply method I these forces are acting entirely upon the continuous phase and again the porosity does not come in. The combined contribution is $-\nabla p \cdot r_0^3$. The body force is easily found to be $\varepsilon \rho_c \mathbf{g} r_0^3$.

The final momentum balance can now be drawn up in a similar way as for the dispersed solid phase:

$$\varepsilon \rho_c \frac{D}{Dt} \mathbf{v}_c = -\nabla \cdot \mathbf{R}_c - \nabla \cdot \tau_c - \nabla p + \varepsilon \rho_c \mathbf{g} + \mathbf{F}_{dc} \qquad (5.14)$$

After substitution of $F_{dc} = (1 - \varepsilon)(\nabla p + \nabla \cdot \tau_c) + F_s$ it is found that

$$\varepsilon \rho_c \frac{D}{Dt} v_c = - \nabla \cdot R_c - \varepsilon(\nabla \cdot \tau_c + \nabla p) + \varepsilon \rho_c g + F_s \qquad (5.15)$$

where

$$R_c = \overline{\varepsilon \rho_c v_c'' v_c''}.$$

5.5 CONSTITUTIVE EQUATIONS

In order to make the momentum equations operational we need information to interpret τ_c, R_c, R_d, σ and F_s in terms of physical properties of the two phases, i.e. viscosity, density and particle diameter on the one hand and their relation with the hydrodynamic variables v_c, v_d (or v_s) and the porosity on the other.

We shall limit this discussion to the case of stationary flow of solid particles and a non-compressible continuous phase which behaves as a Newtonian fluid. This is justified as the pressure differences that occur in most cases of powder handling are so low that the compressibility of the gas phase can be neglected.

When these restrictions are not permitted this has consequences for the correct interpretation of τ_c and F_s.

5.5.1 Re τ_c, R_c and R_d

There is general agreement that the elaboration of τ_c should look similar to that in the case of single-phase flow. For the case of a Newtonian fluid it then holds that

$$\tau_{xx} = -2\mu \partial v_x / \partial x$$
$$\tau_{yy} = -2\mu \partial v_y / \partial y$$
$$\tau_{zz} = -2\mu \partial v_z / \partial z$$
$$\tau_{xy} = \tau_{yx} = -\mu(\partial v_x / \partial y + \partial v_y / \partial x)$$
$$\tau_{xz} = \tau_{zx} = -\mu(\partial v_x / \partial z + \partial v_z / \partial x)$$
$$\tau_{yz} = \tau_{zy} = -\mu(\partial v_y / \partial z + \partial v_z / \partial y)$$

This is often also presented as

$$\tau = -\mu(\nabla v + \nabla v^T) \qquad (5.16)$$

in which ∇v^T is the transposed tensor of ∇v.

As discussed already in connection with the derivation of the convective momentum transport of the dispersed phase, the tensor R_c is due to the tortuous flow of the continuous phase around the solid particles and it has a similar effect on the motion as the tensor τ_c, viz. an exchange of momentum between adjacent streamlines. Hence, it can be accounted for by the same equations as (5.16) but now with a 'turbulent' viscosity μ_t. When we assume that the turbulence is isotropic its effect can easily be evaluated by comparing the corresponding components of τ and R_c:

$$\frac{R_{xx}}{\tau_{xx}} = \frac{\mu_t}{\mu} = \frac{\varepsilon \rho_c v''_{cx} v''_{cx}}{2\mu dv_{cx}/dx} = \frac{\varepsilon \rho_c C |v''_{cx}| \cdot |v''_{cx}|}{2\mu dv_{cx}/dx}$$

in which C is a correlation coefficient < 1 (and in most cases < 0.5.) With $\varepsilon = 0.5$, $\rho = 1\,\text{kg m}^{-3}$, $v_{cx} < 0.02\,\text{m s}^{-1}$ (in the case of fine powders fluidized homogeneously), $d_p < 10^{-4}\,\text{m}$ and $\mu = 2 \times 10^{-5}$ N sm^{-2}, it follows that $\mu_t/\mu \ll 10^{-3}$ where it has been assumed that $v''_{cx} < v_{cx}$ while dv_{cx}/dx has been approximated by $v_{cx}d_p$. Hence in nearly all cases the action of the 'Reynolds stress' in fine powders can be neglected.

With regard to R_d, due to the low mobility of fine powders at not too high a porosity, the 'turbulence' of the solid particles will be very low and hence the 'particle Reynolds stress' can be neglected as well.

5.5.2 Re F_s

The slip force F_s is one of the most important parameters in the analysis of the motion of fine powders. It originates from the velocity difference between the continuous phase and the dispersed phase.

In close packings the pores between the dispersed particles are of the same order of magnitude as the particles themselves. As at the surface of the particles the velocity of the continuous gas phase must be the same as that of the particles, this means that the velocity gradient of the gas phase near the particle surface is very high and proportional to the ratio of the slip velocity v_s to the particle diameter d_p.

As in fine powders the gas flow through the pores generally is at a Reynolds number $\ll 1$, the slip force can be expected to be proportional to

$$(\mu v_s/d_p) \times (\text{specific surface area})$$

and will further depend strongly on the porosity. For a further derivation of the slip force we shall analyse a special case in which:

(1) the flow of both phases is vertical and stationary;
(2) the velocity profile of both phases is flat.

These assumptions hold in many cases of powder flow. In these cases $(D/Dt)v_d$, $(D/Dt)v_c$ and $\nabla \cdot \tau_c$ are all zero.

The momentum equations now reduce to

$$-\varepsilon \nabla p + \varepsilon \rho_c \mathbf{g} + \mathbf{F}_s = 0 \qquad (5.17)$$

for the continuous phase, and

$$-\nabla \cdot \boldsymbol{\sigma} - (1-\varepsilon)\nabla p + (1-\varepsilon)\rho_d \mathbf{g} - \mathbf{F}_s = 0 \qquad (5.18)$$

for the dispersed phase. After elimination of the pressure gradient we obtain

$$\varepsilon \nabla \cdot \boldsymbol{\sigma} = (1-\varepsilon)(\rho_d - \bar{\rho})\mathbf{g} - \mathbf{F}_s \qquad (5.19)$$

in which we have introduced the average density $\bar{\rho}$ of the two-phase system, $\bar{\rho} = \varepsilon \rho_c + (1-\varepsilon)\rho_d$. As reasoned above we may put

$$\mathbf{F}_s = (\mu v_s/d_p) \times (\text{specific surface area}) \times h(\varepsilon) \qquad (5.20)$$

in which $h(\varepsilon)$ is a function of the porosity not yet defined.

We shall now proceed with the case in which $\rho_d > \rho_c$ and the gas flow is upwards and so high that the solids are fluidized. Hence, we need only consider vertical components and may put $\mathbf{g} = (0, 0, -g)$. To a first approximation we may assume that the solid particles are free floating and do not exert forces upon each other, hence $\nabla \cdot \boldsymbol{\sigma} = 0$. Later, in Chapter 7, we shall see that this is not quite true but in our derivation of $\mathbf{F}_s$ this is acceptable. With the specific surface area $S = 6(1-\varepsilon)/d_p$ and with

$$(\rho_d - \bar{\rho}) = \varepsilon(\rho_d - \rho_c)$$

it is found from eqns (5.19) and (5.20) that

$$\mathbf{v}_s = \frac{-d_p^2}{6\mu}(\rho_d - \rho_c)\mathbf{g}\varepsilon/h(\varepsilon)$$

$\mathbf{v}_s$ is generally compared with $\mathbf{v}_{s\infty}$ which is the slip velocity of a single particle at infinite dilution: $\mathbf{v}_s = \mathbf{v}_{s\infty} \cdot f(\varepsilon)$. In the laminar regime $\mathbf{v}_s = (\rho_d - \rho_c)\mathbf{g}d_p^2/18\mu$ according to Stokes.

For the function $f(\varepsilon)$ several proposals have been made, of which we mention the following:

$$f(\varepsilon) = \frac{1}{10}\frac{\varepsilon^2}{(1-\varepsilon)} \qquad \text{Kozeny \& Carman (1937)}$$

$$f(\varepsilon) = \varepsilon^{3 \cdot 6} \qquad \text{Richardson \& Zaki (1954)}$$

$$f(\varepsilon) = \frac{\varepsilon}{\{1 + (1-\varepsilon)^{1/3}\}\exp\dfrac{5(1-\varepsilon)}{3\varepsilon}} \qquad \text{Barnea \& Mizrahi (1973)}$$

In Fig. 5.3 these three relations are compared with each other. The Kozény-Carman equation was originally proposed by Kozény (1927) but further tested and propagated by Carman (1937) In this book it will be referred to as the Carman equation. It must be remarked, however, that this expression cannot be correct at high porosities, since for $\varepsilon \to 1$ this expression goes to infinity and it will be clear that the limit of $f(\varepsilon)$ as $\varepsilon \to 1$ must be 1. Nevertheless, for $\varepsilon < 0 \cdot 7$ the Carman expression gives the best results yet and it will be used further in this book.

The expression for the slip velocity is now found to be

$$\mathbf{F}_s = \frac{180\mu}{d_p^2}\frac{(1-\varepsilon)^2}{\varepsilon}(\mathbf{v}_d - \mathbf{v}_c) \qquad (5.21)$$

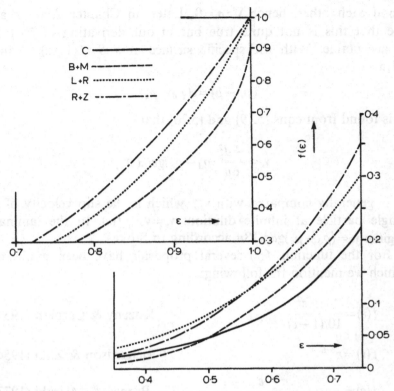

Fig. 5.3. Comparison of various proposals for the porosity function $f(\epsilon)$. C= Carman (1937), B+M=Barnea & Mizrahi (1973), L+R=Loeffler & Ruth (1959), R+Z=Richardson & Zaki (1954).

5.5.3 Re σ

σ is determined by the rheology of the powder and will be treated separately in Chapter 9.

5.6 THE BUOYANCY

In the framework of this book belongs also a discussion on the buoyancy (also called the Archimedes force) which a submerged object experiences from the surrounding medium. According to Archimedes' law this buoyancy is equal to the weight of the surrounding fluid displaced by the submerged object:

$$B = -\rho_m g V_0$$

in which ρ_m is the density of the surrounding fluid and V_0 the volume of the object.

It must be clear that the buoyancy of a submerged object must be due to the variation of the pressure in the surrounding medium around the object. When this pressure is integrated over the surface of the object this results in the buoyancy as formulated by Archimedes:

$$B = \int_{S_0} p \cdot dS_0 = -\int_{V_0} \nabla p \cdot dV_0$$

With ∇p constant and equal to $\rho_m g$ this gives Archimedes' law.

Nevertheless, this formulation leaves open some questions, especially where the surrounding medium is not a pure fluid but a suspension of fine particles in a liquid or a gas.

(a) Does Archimedes' law also hold for the suspension particles themselves?
(b) Which density should then be taken, that of the pure fluid or that of the suspension?
(c) Is it independent of the relative motion of the object?

A closer inspection of the momentum equation (5.13) of the dispersed phase can help us to find the answers to these questions. It should be realized that, of course, the buoyancy must be determined by one of the terms of the momentum equation. In the derivation of this equation (and also of the momentum equation of the continuous phase) all terms are defined as forces or fluxes per volume element of dispersed system (dispersed phase + continuous phase). Hence by division of all terms by a factor of $(1-\varepsilon)$ and subsequent multiplication by the volume of the submerged object, the momentum equation of this object is obtained:

$$(\rho_d V_0) \frac{D}{Dt} v_d = \left\{ -\frac{\nabla \cdot \sigma}{(1-\varepsilon)} - \nabla p - \nabla \cdot \tau_c + \rho_d g - \frac{F_s}{(1-\varepsilon)} \right\} V_0 \quad (5.22)$$

and we recognize the second term on the right-hand side as the buoyancy.

In the interpretation of this equation the following two things should be kept in mind.

(a) The variables occurring in this equation are all averaged variables and not local variables as follows from the derivation of the original equation (5.13). Especially at the surface of the solid particles the average pressure p in the continuous phase can differ considerably from the local pressure, this difference being strongly influenced by the slip between the two participating phases.

(b) In the derivation of eqn (5.13) no assumption has been made as to the size of the particles except that this size should be much smaller than any size of the flow field. Hence eqn (5.22) holds for any particle which is not too large and so also for the suspended particles themselves or for even smaller particles.

We shall now return to the special case of stationary vertical flow without velocity gradients as determined by eqns (5.17)–(5.19). When there is no flow at all $\mathbf{F_s}$ must be zero and from eqn (5.17) it follows that in that case

$$\nabla p = \rho_c \mathbf{g}$$

This means that now in the formulation of Archimedes the density of the pure fluid should be taken for ρ_m as was meant by Archimedes.

On the other hand when there is an upward flow of the continuous phase that is sufficiently strong for the solid particles to be fluidized, $\nabla \cdot \boldsymbol{\sigma} = 0$ and

$$\mathbf{F_s} = (1 - \varepsilon)(\rho_d - \bar{\rho})\mathbf{g}$$

as follows from eqn (5.19).

As $\mathbf{F_s}$ is the slip force per unit volume of dispersed system, $\mathbf{F_s}/(1 - \varepsilon)$ is the slip force per unit volume of dispersed phase and, hence, the slip force $\mathbf{F_p}$ per dispersed particle is

$$F_p = (\rho_d - \bar{\rho})\mathbf{g}V_p \qquad (5.23)$$

As the slip force now makes equilibrium with the two other forces operating on the particle, viz. the gravity $(= \rho_d \mathbf{g} V_p)$ and the buoyancy $\mathbf{B}$, it follows that

$$\mathbf{B} = \bar{\rho} \mathbf{g} V_p$$

and the formulation of Archimedes would give a wrong result. The more general and basic formulation of the buoyancy is given by

$$\mathbf{B} = (\nabla p) \cdot V_p \qquad (5.24)$$

As the pressure gradient need not necessarily be vertical this formulation has the consequence that the buoyancy too need not be vertical.

A proof that the original formulation by Archimedes cannot be right is found in experiments by Van Duyn and Rietema (1982) on liquid fluidization of a binary solid mixture:

(1) ilmenite, particle size 125–250 m^{-6}, density 4300 kg m^{-3}, and
(2) glass beads, particle size 240–280 m^{-6}, density 2500 kg m^{-3}.

The fluidization liquid was water. It was found that at low superficial liquid velocities the ilmenite collected at the bottom of the fluidization apparatus and the glass beads at the top, while at high liquid velocities the reverse was true (see Fig. 5.4).

Where α_1 is the volume fraction of solid 1 and α_2 that of solid 2, then in the stationary state of fluidization:

$$-\alpha_1(dp/dz) - \alpha_1 \rho_{d1} g - F_{s1} = 0$$
$$-\alpha_2(dp/dz) - \alpha_2 \rho_{d2} g - F_{s2} = 0$$
$$-\varepsilon(dp/dz) - \varepsilon \rho_c g + F_{s1} + F_{s2} = 0$$

or by summation

$$- dp/dz - \bar{\rho} g = 0$$

where $\varepsilon = 1 - \alpha_1 - \alpha_2$ and $\bar{\rho} = \varepsilon \rho_c + \alpha_1 \rho_{d1} + \alpha_2 \rho_{d2}$.

The relation between the slip force $\mathbf{F}_{si}$ and the slip velocity v_{si} can be simply expressed by

$$\mathbf{F}_{si} = \mathbf{v}_{si} / m_i$$

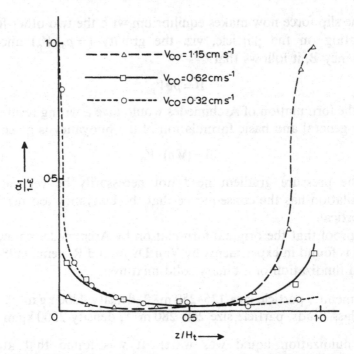

Fig. 5.4. Results of liquid fluidization of a binary solid mixture (experiments of Van Duyn & Rietema, 1982). The fraction of solids occupied by ilmenite as a function of the relative bed height.

in which m_i is the so-called mobility of the solid i, which depends on the particle size, the viscosity and the porosity. With eqns (5.19), (5.20) and the Carman equation for $f(\varepsilon)$ it follows that

$$m_i = \frac{d_{pi}^2 \varepsilon}{180\,\mu(1-\varepsilon)^2} \qquad (5.25)$$

For solid 1 it now holds that

$$\rho_{d1} = \bar{\rho} - \frac{v_{d1}}{\alpha_1 m_1 g} + \frac{v_c}{\alpha_1 m_1 g}$$

and for solid 2

$$\rho_{d2} = \bar{\rho} - \frac{v_{d2}}{\alpha_2 m_2 g} + \frac{v_c}{\alpha_2 m_2 g}$$

At low velocities of the fluidization liquid, solid 1 (ilmenite) collects at the bottom, hence $v_{d1} < 0$ and $v_{d2} > 0$. Hence, in this case

$$\rho_{d1} > \bar{\rho}_l + \left(\frac{v_c}{\alpha_1 m_1 g}\right)_l$$

and

$$\rho_{d2} > \bar{\rho}_l + \left(\frac{v_c}{\alpha_2 m_2 g}\right)_l$$

Similarly, at high liquid velocity

$$\rho_{d1} > \bar{\rho}_h + \left(\frac{v_c}{\alpha_1 m_1 g}\right)_h$$

and

$$\rho_{d2} > \bar{\rho}_h + \left(\frac{v_c}{\alpha_2 m_2 g}\right)_h$$

If the buoyancy were given by $\rho_c g V_p$ in the above equations, both $\bar{\rho}_l$ and $\bar{\rho}_h$ should be replaced by ρ_c. As in all circumstances $\rho_c < \rho_{d1}$ and $\rho_c < \rho_{d2}$ this would be impossible.

REFERENCES

Barnea, E. & Mizrahi, J. (1973). A generalized approach to the fluid dynamics of particulate systems. *The Chem. Engr. J.*, **5**, 171.

Carman, P.C. (1937). Fluid flow through granular beds. *Trans. Inst. Chem. Engrs*, **15**, 150.

Kozény, J. (1927). Über kapillare leitung des Wassers im Boden. *Ber. Wien. Akad.*, **136A**, 291.

Loeffler, A.L. & Ruth, B.F. (1959). Particulate fluidization and sedimentation of spheres. *A.I.Ch.E.J.*, **5**, 310.

Richardson, J.F. & Zaki, W.N. (1954). Sedimentation and fluidization. *Trans. Inst. Chem. Engrs*, **32**, 35.

Rietema, K. & Van den Akker, H.E.A. (1983). On the momentum equations in dispersed two-phase systems. *Int. J. Multiphase Flow*, **9**, 21.

Van Duyn, G. & Rietema, K. (1982). Segregation of liquid fluidized beds. *Chem. Eng. Sci.*, **37**, 727.

6

Stability and Perturbations

NOTATION

a	Growth constant of a perturbation (s^{-1})
A	$180\mu/d_p^2$ $(N\,s\,m^{-4})$
C	$(1-\varepsilon^0)A/(\varepsilon^0)^3\rho_d$ (s^{-1})
d_p	Particle size (m)
E	Elasticity modulus $(N\,m^{-2})$
F_s	Slip force $(N\,m^{-3})$
g	Gravitational acceleration $(m\,s^{-2})$
h	Vertical coordinate (m)
H	Bed height (m)
N_F	Fluidization number (—)
p	Gas pressure $(N\,m^{-2})$
t	Time coordinate (s)
t_s	Time of bed expansion (s)
U	Velocity of a perturbation $(m\,s^{-1})$
U_c	Velocity of a continuity wave $(m\,s^{-1})$
U_{cs}	Velocity of a shock wave $(m\,s^{-1})$
U_d	Velocity of a dynamic wave $(m\,s^{-1})$
v_c	Linear velocity of gas $(m\,s^{-1})$
v_{co}	Superficial gas velocity $(m\,s^{-1})$

v_d	Velocity of particles (m s^{-1})
v_s	Slip velocity $(v_d - v_c)$ (m s^{-1})
v_t	Superficial velocity of gas and powder together (m s^{-1})
W	Wall friction (N m^{-3})
ε	Porosity (—)
ε_{mb}	Porosity at bubble point (—)
μ	Viscosity of gas (N s m^{-2})
ρ_c	Density of gas phase (kg m^{-3})
ρ_d	Density of solid phase (kg m^{-3})
σ	Shear stress tensor in solid phase (N m^{-2})
τ	Shear stress tensor in gas phase (N m^{-2})
ω	Frequency of a perturbation (s^{-1})

6.1 INTRODUCTION

In this chapter we shall deal with the stability of homogeneous gas-fluidized beds of fine powders and with the evolution of perturbations in such a bed. Later on in this book where we deal with applications, we shall see that this study will lead to a better understanding of the behaviour of fine powders during handling.

As mentioned in Chapter 3 the initial effect of increasing the superficial gas velocity through a homogeneous gas-fluidized bed is expansion of the bed. When the gas velocity exceeds a certain critical value, however, the bed will become unstable: it will segregate into a dense phase with a low porosity and a phase of fast-rising, more or less spherical voids (the so-called 'bubble-phase').

Instability of a system is always initiated by small perturbations introduced from the outside. As long as these perturbations are damped down fast enough (by whatever cause) the system will remain stable. When the perturbations grow, however, the limit of stability is exceeded and the system becomes unstable. In that case the forces that counteract the perturbations are obviously not strong enough.

6.2 CONTINUITY WAVES

A perturbation in a two-phase system manifests itself as a local change in porosity that propagates through the system with a certain velocity. When one phase of the system is stationary and the other phase is flowing, as in a fluidized bed, for example, the direction of propagation coincides with the direction of flow. The velocity of propagation depends on the local porosity.

Such a disturbance is called a continuity wave since its propagation velocity can be derived from the principle of continuity. The propagation velocity U_c is determined by

$$\left(\frac{\partial \varepsilon}{\partial t}\right)_h = -U_c \left(\frac{\partial \varepsilon}{\partial h}\right)_t \qquad (6.1)$$

This can be understood as follows. Suppose the continuity wave moves upwards. At time t the porosity at height h equals ε and at height $(h + \Delta h)$ equals $(\varepsilon + \Delta \varepsilon)$. At time $(t + \Delta t)$ the continuity wave will have moved over a distance Δh so that then at the height $(h + \Delta h)$ the porosity has become equal to ε. Hence,

$$\Delta \varepsilon = \left(\frac{\partial \varepsilon}{\partial h}\right)_t \cdot \Delta h$$

but also

$$\Delta \varepsilon = -\left(\frac{\partial \varepsilon}{\partial t}\right)_h \cdot \Delta t$$

The propagation velocity is found from

$$U_c = \frac{\Delta h}{\Delta t} = \frac{-(\partial \varepsilon/\partial t)_h}{(\partial \varepsilon/\partial h)_t} \qquad (6.2)$$

The change of porosity, of course, must be coupled with a small displacement of the dispersed phase. When only vertical velocities occur the continuity equation of the dispersed phase runs:

$$\frac{\partial \varepsilon}{\partial t} = \frac{\partial}{\partial h}\{(1 - \varepsilon)v_d\} = (1 - \varepsilon)\frac{\partial v_d}{\partial h} - v_d \frac{\partial \varepsilon}{\partial h} \qquad (6.3)$$

From eqn (6.2) it now follows that

$$U_c = v_d - (1 - \varepsilon)\frac{\partial v_d}{\partial \varepsilon} \qquad (6.4)$$

As this equation is not very manageable we transform it into an expression of the slip velocity v_s and the porosity. To that end we first introduce the so-called superficial velocities v_{c0} and v_{d0}. These are defined as the total volume of the phase concerned that moves per unit cross-section and per unit time through the bed. Hence, $v_{c0} = \varepsilon v_c$ and $v_{d0} = (1-\varepsilon)v_d$. In the stationary state at any time the total volume flow $v_t = v_{c0} + v_{d0}$ must be the same everywhere in the bed. With $v_s = v_d - v_c$ it follows that also $(v_d - \varepsilon v_s)$ is constant and hence

$$\frac{\partial v_d}{\partial \varepsilon} = \frac{\partial}{\partial \varepsilon}(\varepsilon v_s)$$

When these relations are substituted in eqn (6.4) it is found that

$$U_c = v_t + \varepsilon v_s - (1-\varepsilon)\frac{\partial}{\partial \varepsilon}(\varepsilon v_s)$$

or

$$U_c = v_t - \frac{\partial}{\partial \varepsilon}\{\varepsilon(1-\varepsilon)v_s\} \tag{6.5}$$

With

$$v_s = -\frac{(\rho_d - \rho_c)g d_p^2}{180\,\mu}\left(\frac{\varepsilon^2}{1-\varepsilon}\right)$$

it will be clear that U_c depends strongly on the porosity and increases when ε increases.

In fluidization $v_{d0} = 0$, hence from eqn (6.5) it follows that

$$U_c = (1-\varepsilon)\frac{\partial}{\partial \varepsilon}(v_{c0}) \tag{6.6}$$

and with the expression for $v_s = -v_{c0}/\varepsilon$ it can be derived that in this case

$$U_c = \frac{3-2\varepsilon}{\varepsilon}v_{c0} \tag{6.7}$$

We shall now consider the situation of a homogeneous fluidized bed with everywhere a porosity ε_1 in which we suddenly reduce the gas supply so that after some time a new situation has arisen with everywhere a porosity ε_2. From the moment the gas supply is reduced a slowly rising zone of densification will occur. In this zone the porosity changes continuously from ε_1 at the top to ε_2 at the bottom and hence a whole band of continuity waves moves up. Above this zone the porosity remains at the original value ε_1. Since $\varepsilon_2 < \varepsilon_1$ the lower side of the densification zone will move upwards at a lower velocity than the upper side. Hence, this zone of densification grows in height until its top has reached the top of the bed. Of course, the top of the bed moves down slowly until the lower side of the densification zone has reached the top. At that moment the whole process of densification is finished and the bed will everywhere have reached the final porosity ε_2. Schematically the process of densification is represented in Fig. 6.1.

In the above analysis we have neglected the expansion of the fluidization gas owing to the decrease in pressure when we move up in the bed. At high bed heights—when the pressure drop over the bed cannot be neglected compared to the pressure of the

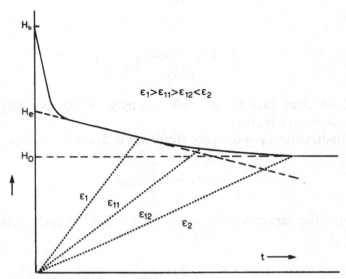

Fig. 6.1. Schematic densification of powder bed after reduction of gas velocity.

fluidization gas at the bottom—the analysis is somewhat more complicated but largely follows the same pattern.

6.3 THE CONTINUITY SHOCK WAVE

Again we consider a homogeneous gas-fluidized bed with porosity ε_1 but now we suddenly increase the gas supply to a value such that finally a homogeneous porosity $\varepsilon_2 > \varepsilon_1$ will be reached. Again the region where the porosity changes from ε_1 to ε_2 moves up. Now, however, the continuity wave corresponding to the new porosity ε_2 will move faster than the continuity wave belonging to the original porosity. The fast continuity wave will try to overtake the slow one but as this is not possible the continuity waves will agglomerate and form a shock wave where the porosity jumps from ε_1 to ε_2.

The derivation of the velocity of a continuity wave does not hold for a shock wave as across the shock the porosity varies discontinuously. A simple derivation of the shock wave velocity now follows from macroscopic considerations.

Assuming that the shock wave velocity is constant and the bed height after expansion is H_2 the total time of expansion $t_s = H_2/U_{cs}$. The extra volume of gas per unit cross-section which the bed adsorbs is $(H_2 - H_1)$ where H_1 is the original bed height. This amount of gas, however, is also equal to $t_s(v_{c02} - v_{c01})$. With $H_2(1 - \varepsilon_2) = H_1(1 - \varepsilon_1)$ and elimination of t_s it is easily found that

$$U_{cs} = \frac{1 - \varepsilon_1}{\varepsilon_2 - \varepsilon_1} (v_{c02} - v_{c01}) \qquad (6.8)$$

Remark 1: As in the situation described above, a bed layer of higher density (lower porosity) is lying above a bed layer of lower density; this situation is not stable and will tend to reverse itself. When the relative increase in gas supply is not too high and the bed diameter is not too large the expansion can proceed more regularly and the moment of inversion can be delayed.

Remark 2: A more complete discussion of continuity waves and shock waves in two-phase systems is to be found in Wallis (1969).

6.4 STABILITY OF A HOMOGENEOUS GAS-FLUIDIZED BED

Fine powders can be fluidized homogeneously by gas, i.e. they display homogeneous expansion at not too high fluidization velocity, while the pressure drop over the bed is almost constant. Depending on the nature of the powder as determined by the particle size, particle density and powder cohesion and on the gas viscosity, bed expansions of 20–50% and sometimes even up to 100% can be reached.

By means of a stability analysis this phenomenon will be further examined. We start from the equations of motion of a two-phase system as derived in Chapter 5.

In order not to complicate matters unnecessarily, the equations will be simplified by neglecting a few minor terms and by confining the analysis to vertical motion only. This latter restriction is justified since we have learned already that in a fluidized bed perturbations always move up.

In the momentum equation of the gas phase we will neglect certain terms as follows.

(1) The shear stress tensor τ_c can be neglected because of the low viscosity of the gas phase and because the velocity profile of the gas phase is almost flat. The slip force F_s, however, which is also proportional to the gas viscosity cannot be neglected because of the steep velocity gradients in the pores between the particles.

(2) As reasoned in Section 5.5.1, R_c can be neglected as well.

(3) The dynamic acceleration term $\varepsilon\rho_c(D/Dt)v_c$ can be neglected because the gas density is low and the variations of the gas velocity are small.

All this means that the momentum equation of the gas phase is reduced to

$$-\varepsilon\frac{\partial p}{\partial h}-\varepsilon\rho_c g+F_s=0 \tag{6.9}$$

In the momentum equation of the dispersed phase we can, of course, also neglect the term $\nabla\cdot\tau_c$ and as indicated in Section 5.5.1 the term $\nabla\cdot R_d$. What remains is to elaborate the term $\nabla\cdot\sigma$. We are again only interested in the vertical component:

$$[\nabla\cdot\sigma]_h=\frac{\partial\sigma_{hh}}{\partial h}+\frac{\partial\sigma_{xh}}{\partial x}+\frac{\partial\sigma_{yh}}{\partial y} \tag{6.10}$$

The last two terms of this equation finally lead to the wall friction after integration over x and y respectively. As the wall friction of an expanded bed is very small we shall represent these two terms together by a small term W.

The term $(\partial/\partial h)\sigma_{hh}$ needs some further discussion and elaboration. As mentioned already in Chapter 3 a powder bed exhibits a certain elasticity, which opposes a possible deformation of the bed such as a small decrease or increase of the local porosity. Therefore, we introduce an elasticity modulus E, measured in $N\,m^{-2}$, to quantify this elasticity. Hence, we put

$$\partial\sigma_{hh}/\partial h=-E\partial\varepsilon/\partial h$$

It should be realized that the elasticity modulus E is characteristic of the packing and is determined by the particle size, the cohesion force between the individual particles and the coordination number (see Section 2.4). It is constant only between the elasticity limits. When these are exceeded the elasticity modulus jumps to a new value; the powder sets at a new structure with fewer contact points (lower coordination number) when the porosity is increased or with more contact points (higher coordination number) when the porosity is decreased.

The final momentum equation of the dispersed phase now reads:

$$(1-\varepsilon)\rho_d\left\{\frac{\partial v_d}{\partial t}+v_d\frac{\partial v_d}{\partial h}\right\}+(1-\varepsilon)\frac{\partial p}{\partial h}+(1-\varepsilon)\rho_d g-E\frac{\partial\varepsilon}{\partial h}+W+F_s=0 \tag{6.11}$$

The elaborated continuity equations run as follows:

$$\frac{\partial \varepsilon}{\partial t} + v_c \frac{\partial \varepsilon}{\partial h} + \varepsilon \frac{\partial v_c}{\partial h} = 0 \tag{6.12}$$

for the gas phase at constant density, and

$$\frac{\partial \varepsilon}{\partial t} + v_d \frac{\partial \varepsilon}{\partial h} - (1-\varepsilon) \frac{\partial v_d}{\partial h} = 0 \tag{6.13}$$

for the dispersed phase. So we have four equations (6.9, 6.11, 6.12 and 6.13) with four variables, viz. p, ε, v_d and v_c. By means of eqn (6.9) we can eliminate the pressure from eqn (6.11) to give

$$(1-\varepsilon)\rho_d \left\{ \frac{\partial v_d}{\partial t} + v_d \frac{\partial v_d}{\partial h} \right\} + W - E \frac{\partial \varepsilon}{\partial h} + (1-\varepsilon)(\rho_d - \rho_c)g + \frac{F}{\varepsilon} = 0 \tag{6.14}$$

The next step is to put the three remaining variables as the sum of a stationary value and a variation or perturbation of this stationary state:

$$\varepsilon = \varepsilon^\circ + \varepsilon'$$
$$v_c = v_c^\circ + v_c'$$
$$v_d = v_d'$$

($v_d = 0$ in the stationary state). There will also be a perturbation in the term (F/ε) of eqn (6.14) which means that we must put

$$\left(\frac{F}{\varepsilon}\right) = \left(\frac{F}{\varepsilon}\right)^\circ + \frac{d}{d\varepsilon}\left(\frac{F}{\varepsilon}\right) \cdot \varepsilon' + \frac{d}{dv_c}\left(\frac{F}{\varepsilon}\right)v_c' + \frac{d}{dv_d}\left(\frac{F}{\varepsilon}\right)v_d' \tag{6.15}$$

Since we want to investigate under what circumstances the bed will become unstable, or in other words each perturbation will grow, these perturbations can be assumed to be still small and hence products of these variations can be neglected.

To elaborate eqn (6.15) we make use of eqn (5.21). We now find, with $A = 180\,\mu/d_p^2$, that

$$(1-\varepsilon^\circ)\rho_d \frac{\partial v_d'}{\partial t} + W - E\frac{\partial \varepsilon'}{\partial h} + (1-\varepsilon^\circ-\varepsilon')(\rho_d-\rho_c)g$$

$$+2\frac{(1-\varepsilon^\circ)}{(\varepsilon^\circ)^3}Av_c^\circ\varepsilon' + A\left(\frac{1-\varepsilon^\circ}{\varepsilon^\circ}\right)^2(v_d'-v_c'-v_c^\circ) = 0 \qquad (6.16)$$

For the stationary state,

$$W + (1-\varepsilon^\circ)(\rho_d-\rho_c)g + \left(\frac{F}{\varepsilon}\right)^\circ = 0 \qquad (6.17)$$

Subtracting eqn (6.17) from eqn (6.16) gives

$$(1-\varepsilon^\circ)\rho_d\frac{\partial v_d'}{\partial t} - E\frac{\partial \varepsilon'}{\partial h} - \varepsilon'(\rho_d-\rho_c)g$$

$$+2\frac{1-\varepsilon^\circ}{(\varepsilon^\circ)^3}Av_c^\circ\varepsilon' + A\left(\frac{1-\varepsilon^\circ}{\varepsilon^\circ}\right)^2(v_d'-v_c') = 0 \qquad (6.18)$$

The same assumptions introduced in the continuity equations lead for the gas phase to

$$\varepsilon^\circ\frac{\partial v_c'}{\partial h} + v_c^\circ\frac{\partial \varepsilon'}{\partial h} + \frac{\partial \varepsilon'}{\partial t} = 0 \qquad (6.19)$$

and for the dispersed phase to

$$(1-\varepsilon^\circ)\frac{\partial v_d'}{\partial h} - \frac{\partial \varepsilon'}{\partial t} = 0 \qquad (6.20)$$

When we differentiate eqn (6.18) with respect to h and eqn (6.20) with respect to t, while also applying eqn (6.19), v_d and v_c can be eliminated from eqn (6.18). This yields an equation exclusively of ε':

$$\rho_d\frac{\partial^2\varepsilon'}{\partial t^2} - E\frac{\partial^2\varepsilon'}{\partial h^2} - (\rho_d-\rho_c)g\frac{\partial \varepsilon'}{\partial h} + \frac{1-\varepsilon^\circ}{(\varepsilon^\circ)^3}A\left\{(3-\varepsilon^\circ)v_c^\circ\frac{\partial \varepsilon'}{\partial h} + \frac{\partial \varepsilon'}{\partial t}\right\} = 0 \qquad (6.21)$$

From eqns (6.17) and (5.21), neglecting the term W, it can be derived that

$$(\rho_d - \rho_c)g = Av_c^\circ \frac{1-\varepsilon^\circ}{(\varepsilon^\circ)^2}$$

With this result, eqn (6.21) can be further simplified. After dividing by ρ_d we find

$$\frac{\partial^2 \varepsilon^1}{\partial t^2} - \frac{E}{\rho_d}\frac{\partial^2 \varepsilon'}{\partial h^2} + \frac{1-\varepsilon^\circ}{(\varepsilon^\circ)^3}\frac{A}{\rho_d}\left\{(3-2\varepsilon^\circ)v_c^\circ\frac{\partial \varepsilon'}{\partial h} + \frac{\partial \varepsilon'}{\partial t}\right\} = 0 \quad (6.22)$$

The first two terms of this equation together form the wave equation of a dynamic wave with propagation velocity $U_d = \sqrt{E/\rho_d}$. In the third term the expression $(3-2\varepsilon^\circ)v_c^\circ$ has been mentioned before as the propagation velocity U_c of a continuity wave. With $C = (1-\varepsilon^\circ)A/\rho_d(\varepsilon^\circ)^3$, eqn (6.22) can now be rewritten:

$$\frac{\partial^2 \varepsilon'}{\partial t^2} - U_d^2\frac{\partial^2 \varepsilon'}{\partial h^2} + C\left(U_c\frac{\partial \varepsilon'}{\partial h} + \frac{\partial \varepsilon'}{\partial t}\right) = 0 \quad (6.23)$$

We now come to the last step in which we suppose that any perturbation can be described by a Fourier series. If we can show that for any term of such a series the bed will be stable then probably we have found the criterion for bed stability. The general shape of a term from the Fourier series runs:

$$\varepsilon' = e^{at + i\omega(t - h/U)} \quad (6.24)$$

Here U is the propagation velocity of the disturbance and ω its frequency; e^{at} is the amplitude of the disturbance. If $a > 0$ the disturbance will grow and if $a < 0$ it will die out regardless of the frequency and the propagation velocity.

Substituting eqn (6.24) into eqn (6.23) gives

$$(a + i\omega)^2 - \left(\frac{i\omega}{U}\right)^2 + C\left(-i\omega\frac{U_c}{U} + a + i\omega\right) = 0 \quad (6.25)$$

Both the real part and the imaginary part must be zero. For the imaginary part this gives

$$a = \frac{C}{2}\left(\frac{U_c}{U} - 1\right) \quad (6.26)$$

and for the real part

$$\omega^2 = \frac{a(a+C)}{1-\left(\dfrac{U_d}{U}\right)^2} = \frac{C^2}{4}\left(\frac{U_c^2-U^2}{U^2-U_d^2}\right) \qquad (6.27)$$

Naturally ω must be real and, hence, $\omega^2 > 0$. This is the case when:

$$U_c^2 > U > U_d^2$$

or $\qquad\qquad U_c^2 < U < U_d^2 \qquad\qquad (6.28)$

In the first case the constant a will be >0 as follows from eqn (6.26). Hence the perturbations will grow and the bed is unstable. In the second case $a<0$ and the perturbations die out. The necessary condition for stability therefore is that

$$U_d > U_c$$

When in a stable bed the superficial gas velocity is increased the bed will further expand and the porosity will increase. Hence, the continuity wave velocity

$$U_c = v_c(3-2\varepsilon) = \frac{d_p^2}{180\mu}(\rho_d-\rho_c)g\frac{\varepsilon^2}{(1-\varepsilon)}(3-2\varepsilon) \qquad (6.29)$$

will increase. Since the elasticity modulus will decrease with increasing porosity (decrease of the coordination number), the dynamic wave velocity U_d will decrease, and at a critical porosity ε_{mb} the situation will be reached where U_d becomes equal to U_c, which therefore is the limit of the stability range. By measuring experimentally this critical porosity ε_{mb} and the corresponding gas velocity $v_{c,mb}$ the elasticity modulus E can be calculated from $U_d = U_c$ or from

$$E = \rho_d\{v_{c,mb}(3-2\varepsilon_{mb})\}^2 \qquad (6.30)$$

It should be realized that the value of E thus determined only pertains at the corresponding porosity ε_{mb} and only for the powder investigated. To determine the value of E at other porosities an obvious method is to fluidize the powder bed with various gases with different viscosities. The maximum homogeneous bed expansion is then found at a different porosity. Hence, different values of E pertaining to different porosities are found.

When $U_d < U_c$ the perturbations will grow. However, how this growth takes place exactly cannot be derived from this stability theory since in that case it is no longer valid to neglect the products of variations in ε, v_c and v_d. It is generally assumed, however, that the growing perturbations will finally develop in fast-rising bubbles.

When, for a certain powder fluidized with a gas of low viscosity (e.g. hydrogen or methane), even at the incipient fluidization velocity $U_d < U_c$ homogeneous fluidization is not possible, the powder then seems to be a so-called B-powder as defined by Geldart (see Section 1.6). Nevertheless, it may be possible for the same powder to be fluidized homogeneously with a gas of higher viscosity (e.g. argon or neon), so that it now seems to be a so-called A-powder. For this reason it is better not to distinguish between A-, B- and C-powders as proposed by Geldart but between A-, B- and C-behaviour of the powders.

The criterion $U_d > U_c$ can also be elaborated in the dimensionless form:

$$N_F = \frac{\rho_d(\rho_d - \rho_c)^2 g^2 d_p^4}{\mu^2 E} < \left\{\frac{180(1-\varepsilon)}{\varepsilon^2(3-2\varepsilon)}\right\}^2$$

where N_F is called the fluidization number.

6.5 OTHER THEORIES

Many investigators have tackled the problem of stability of homogeneous gas–solid fluidization or as some investigators call it 'the problem of delayed bubbling'. The majority (e.g. Verloop & Heertjes, 1970; Oltrogge, 1972; Foscolo and Gibilaro, 1984) recognize the necessity that the dynamic wave velocity U_d should be larger than or equal to the continuity wave velocity U_c.

However, they deny the existence or the importance of interparticle forces and seek the origin of the dynamic wave in hydrodynamics. This is remarkable since from the momentum equation (6.11), though without the elasticity term, no elastic force can be derived.

In the derivation of their theory Foscolo and Gibilaro do a mental experiment with a movable frictionless piston as gas distributor. When this piston is moved up a little by means of an external force the bed at the bottom is compressed and subsequently this compression wave moves up through the bed. They state that this is a dynamic wave with a propagation velocity $U_d = (\partial p/\partial \rho)^{0.5}$. This statement is not correct. In fact it is just a perturbation induced by an external force and not the consequence of an intrinsic property of the system. Unlike elastic waves this compression wave does not move in two directions (up and down) but only upwards. Hence, it is just a continuity wave. Furthermore, the expression which they apply for the dynamic wave velocity does not hold for a dispersed two-phase system such as a fluidized bed but only for a homogeneous compressible one-phase system.

In a recent paper by Rietema and Piepers (1990) the theory of Foscolo and Gibilaro is further criticized. It is shown that they also made a serious mistake (neglecting a minus sign) in their calculations.

All theories referred to above can be expressed by the simple relation

$$Ar < f(\varepsilon)$$

in which Ar is the Archimedes number, given by $(\rho_d - \rho_c) \, g d_p^3/\mu^2$. The mutual differences between these theories are to be found only in the dependence on the porosity.

6.6 CHANNELLING

As mentioned in Section 1.6, some powders show the phenomenon of channelling (C-behaviour). The stability theory as developed in this chapter does not give insight into this phenomenon nor does it lead to a criterion that must be exceeded to result in C-behaviour of the powder.

At present no theory on channelling is available. According to

Geldart it is only certain particular powders with a very small average particle size and a high particle density which show channelling. In its generality this is not correct. As an example we mention potato starch (see Fig. 1.1 in Chapter 1) whose average particle size and particle density are similar to those of cracking catalyst; while cracking catalyst can be fluidized very well and shows high bed expansion, potato starch shows the phenomenon of channelling very seriously and gives hardly any bed expansion.

It is remarkable that especially soft powders such as potato starch, flour, cement, iron oxide and limestone give rise to channelling. It is also remarkable that all these powders are very cohesive. We believe that it is the ratio between the elasticity modulus and the cohesion which determines whether a powder will give rise to channelling or not.

In Chapter 8 a theoretical model on powder structures will be developed from which it follows indeed that the above ratio decreases when the hardness of the powder decreases.

REFERENCES

Foscolo, P.U. & Gibilaro, L.G. (1984). A fully predictive criterion for the transition between particulate and aggregative fluidization. *Chem. Eng. Sci.*, **39**, 1667.

Oltrogge, R.D. (1972). *Gas fluidized beds of fine particles*. PhD Thesis, University of Michigan.

Rietema, K. & Piepers, H.W. (1990). The effect of interparticle forces on the stability of gas fluidized beds. I. Experimental evidence. *Chem. Eng. Sci.*, **45**, 1627.

Verloop, J. & Heertjes, P.M. (1970). Shock waves as a criterion for the transition from homogeneous to heterogeneous fluidization. *Chem. Eng. Sci.*, **25**, 825.

Wallis, G.B. (1969). *One-dimensional Two-phase Flow*. McGraw-Hill, New York.

7

Experiments with Homogeneous Fluidization

NOTATION

C	See Chapter 6
d_p	Particle size (m)
E	Elasticity modulus ($N\,m^{-2}$)
E_{mb}	Value of E at the bubble point ($N\,m^{-2}$)
E_{50}	Value of E at a porosity of 50% ($N\,m^{-2}$)
g	Gravitational acceleration ($m\,s^{-2}$)
H_{mb}	Bed height at the bubble point (m)
N_F	Fluidization number (—)
r	Radius of rotation (m)
U_c	Velocity of continuity wave ($m\,s^{-1}$)
v_{co}	Superficial gas velocity ($m\,s^{-1}$)
v_{mb}	Gas velocity at the bubble point ($m\,s^{-1}$)
v_{mf}	Gas velocity at incipient fluidization ($m\,s^{-1}$)
ε	Porosity (—)
ε_{mb}	Porosity at the bubble point (—)
μ	Gas viscosity ($N\,s\,m^{-2}$)
ρ_c	Density of gas ($kg\,m^{-3}$)
ρ_d	Density of solid particles ($kg\,m^{-3}$)
ω	Angular velocity (s^{-1})

7.1 INTRODUCTION

If the theory presented in the foregoing chapter is correct, the elasticity modulus E must be a property of the powder packing. Unfortunately no method is known of measuring this property directly. A first attempt in this direction (Mutsers & Rietema, 1979) was unsuccessful. Therefore, corroboration of the above statement was sought by correlating experimentally derived values of E with other powder properties such as porosity, average particle size, particle size distribution and particle hardness.

This chapter describes experiments to measure E under various conditions of gas viscosity and gas pressure. Also experiments were carried out in a fluidized bed mounted in a large, so-called human centrifuge to investigate the effect of gravity.

7.2 DESCRIPTION OF APPARATUS

The apparatus in which the experiments are carried out is schematically represented in Fig. 7.1. The fluidized bed proper consisted of a 1450 mm high glass tube of 82 mm i.d. A porous brass plate was used as a gas distributor. The fluidized bed was part of a closed loop system through which the fluidizing gas—initially fed from a pressurized gas cylinder—was circulated by means of a gas pump. Buffering vessels before and after the pump reduced flow rate fluctuations. A heat exchanger fed with cooling water kept the temperature of the whole system constant at 18°C. The gas flow rate in the system was measured with a set of calibrated rotameters. The height of the bed was measured by reference to a calibrated tape on the outside of the glass tube. The whole system was prepared for experiments with different gases at pressures up to 15 bar. When changing over from one gas to another the whole equipment was evacuated three times with subsequent refreshing of the gas phase.

Two powders were investigated, viz. fresh cracking catalyst (FCC) and polypropylene (PP). The physical properties of these

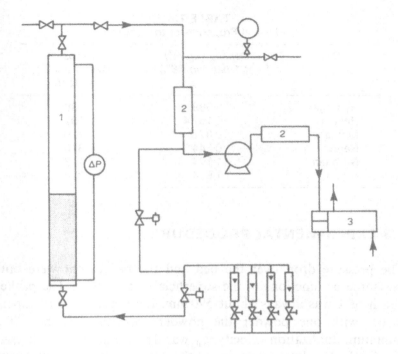

Fig. 7.1. Experimental setup: 1 = fluidized bed, 2 = buffering vessel, 3 = heat exchanger.

TABLE 7.1
Physical Properties of the Powders

Powder[a]	Skeletal density ρ_s (kg/m³)	Internal porosity ϵ_i	Particle density ρ_p (kg/m³)	Surface–volume mean diameter d_p (µm)
FCC	2 465	0·640	887	59·4
PP	918	0·389	561	76·0

[a]FCC = fresh cracking catalyst, PP = polypropylene.

powders are given in Table 7.1. Experiments were carried out with six different gases: hydrogen, helium, methane, neon, nitrogen and argon. The densities of these gases at 1 bar and 18°C as well as their viscosities are given in Table 7.2.

TABLE 7.2
Physical Properties of the Gases

Gas	Density (kg/m³) at 1 bar and 18°C	Viscosity (Ns/m² × 10⁷) at 18°C
Hydrogen	0·089	88
Helium	0·167 4	192
Methane	0·672	110
Neon	0·844	310
Nitrogen	1·174	173
Argon	1·674	220

7.3 EXPERIMENTAL PROCEDURE

The pressure drop over the bed and the bed height were both measured as functions of the superficial gas velocity. The packed bed height was always about 590 mm. During a series of experiments with one powder the powder was not refreshed. The minimum fluidization velocity v_{mf} was determined by the intersection of the two linear portions of the pressure-drop velocity curve (see Fig. 7.2). The measured values of v_{mf}, within the experimental error, did not show any systematic dependence on pressure. The results could very well be represented by Ergun's equation.

The measurement of the maximum stable bed height H_{mb} and of the bubble point velocity v_{mb}, i.e. the superficial gas velocity at which the first small bubbles appear, calls for more explanation.

When v_{mb} is just exceeded the bed becomes unstable, which manifests itself by slow rising and falling of the bed height. During the rising phase no bubbles are produced. When the maximum bed height is reached, however, a train of bubbles is suddenly formed near the gas distributor. These bubbles rise fast and grow in size by extraction of gas from the dense phase which results in a dense phase of lower porosity once all bubbles have left the bed; hence the bed height goes down fast. When the lower bed height is reached the whole process repeats itself. When v_{mb} is further exceeded the oscillation frequency increases; more gas is extracted from the dense phase while the fluctuation amplitude also in-

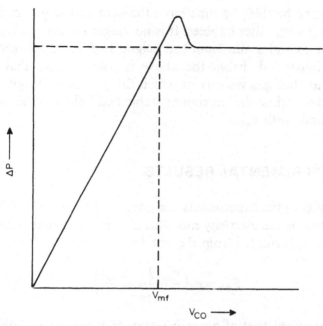

Fig. 7.2. Pressure drop over bed versus the superficial gas velocity v_{co}; definition of v_{mf}.

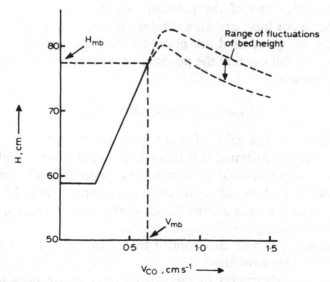

Fig. 7.3. Bed height and bed height fluctuation; definitions of H_{mb} and v_{mb}.

creases. The bubbles become more frequent and larger, mainly by coalescing with other bubbles. It is no longer possible to recognize a maximum and a minimum bed height. In Fig. 7.3 this process is further illustrated. From the above it will be clear that v_{mb} is defined as that gas velocity at which the process of height fluctuation starts, while the maximum stable bed height is the one that corresponds with v_{mb}.

7.4 EXPERIMENTAL RESULTS

The results of the experiments are given in Tables 7.3 and 7.4. For all conditions the elasticity modulus E at the maximum stable bed height was calculated from the relation

$$E_{mb} = \rho_d \left(\frac{v_{mb}(3 - 2\varepsilon_{mb})}{\varepsilon_{mb}} \right)^2 \qquad (7.1)$$

If it is assumed that at a gas pressure of 1 bar adsorption effects on the interparticle forces can still be neglected (see Chapter 4), E_{mb} at this pressure should be a function only of the nature of the solid particles and of the porosity. To determine this porosity dependence for fresh cracking catalyst, the E_{mb} values at 1 bar for all of the six gases used are plotted versus ε_{mb} in Fig. 7.4. From this figure it follows that the porosity dependence can be described by the relation

$$E_{mb} = E_{50} \exp[15(0·50 - \varepsilon_{mb})] \qquad (7.2)$$

in which E_{50} is the value of E at a porosity of 0·50.

If it is further assumed that this relation holds also at higher gas pressures, it can be used to calculate E_{50} values for all conditions and thus to find the effect of gas adsorption. In Fig. 7.5 E_{50} is plotted versus the gas density for all of the six gases and at all gas pressures, which gives a unique relation.

Equation (7.2) was also applied to derive E_{50} values for polypropylene. This gave another but also unique correlation, suggesting that gas adsorption to this solid is less strong than to fresh cracking catalyst (see also Fig. 7.5).

TABLE 7.3

Results of Experiments on Homogeneous Fluidization of Fresh Cracking Catalyst with Different Gases

Pressure (bar)	ρ_c (kg m^{-3})	u_{mb} (cm/s)	H_{mb} (cm)	ϵ_{mb}	E_{mb} (N/m^2)	E_{50} (N/m^2)
Hydrogen						
1	0·084 3	0·80	71·2	0·511	0·850	1·00
3	0·253	0·72	71·2	0·511	0·688	0·81
6	0·506	0·72	71·2	0·511	0·688	0·81
9	0·759	0·72	71·2	0·511	0·688	0·81
12	1·012	0·72	72·0	0·517	0·665	0·86
15	1·265	0·72	72·2	0·518	0·661	0·87
Helium						
1	0·167 4	0·57	82·5	0·578	0·293	0·944
3	0·502	0·55	83·0	0·580	0·270	0·896
6	1·004	0·55	83·5	0·582	0·267	0·913
9	1·507	0·55	83·5	0·582	0·267	0·913
12	2·01	0·55	83·6	0·583	0·266	0·923
15	2·51	0·56	85·0	0·590	0·264	1·01
Methane						
1	0·672	0·70	75·5	0·538	0·560	0·99
3	2·015	0·73	77·0	0·547	0·573	1·13
6	4·03	0·83	79·3	0·560	0·686	1·69
9	6·05	0·91	81·0	0·570	0·782	2·23
12	8·06	0·94	82·4	0·577	0·802	2·55
15	10·08	0·98	83·9	0·584	0·838	2·95
Neon						
1	0·844	0·50	91·0	0·617	0·182	1·05
3	2·532	0·50	93·0	0·625	0·173	1·13
6	5·064	0·55	96·0	0·637	0·196	1·53
9	7·596	0·58	97·5	0·642	0·213	1·79
12	10·13	0·63	99·8	0·651	0·240	2·31
15	12·66	0·70	104·0	0·664	0·276	3·23
Nitrogen						
1	1·174	0·632	83·7	0·583	0·351	1·22
3	3·522	0·665	85·4	0·591	0·371	1·45
6	7·044	0·790	87·5	0·601	0·495	2·25
9	10·57	0·850	91·2	0·618	0·522	3·06
12	14·09	0·875	92·7	0·624	0·535	3·44
15	17·61	0·950	95·5	0·635	0·594	4·50
Argon						
1	1·674	0·590	87·0	0·600	0·278	1·25
3	5·022	0·650	90·4	0·615	0·310	1·74
6	10·04	0·720	94·5	0·631	0·349	2·50
9	15·07	0·850	98·0	0·644	0·453	3·93
12	20·01	0·940	101·0	0·655	0·522	5·34
15	25·10	1·03	104·0	0·665	0·593	7·04

TABLE 7.4

Results of Experiments on Homogeneous Fluidization of Polypropylene Powder with Different Gases

Pressure (bar)	Nitrogen					Argon				
	U_{mb} (cm s^{-1})	H_{mb} (cm)	ϵ_{mb}	E_{mb} (N/m^2)	E_{50} (N/m^2)	u_{mb} (cm/s)	H_{mb} (cm)	ϵ_{mb}	E_{mb} (N/m^2)	E_{50} (N/m^2)
1	0·58	68·9	0·512	0·278	0·33	0·56	69·5	0·515	0·257	0·32
3	0·62	71·7	0·530	0·289	0·45	0·70	73·0	0·538	0·352	0·62
6	0·72	73·0	0·538	0·367	0·65	0·79	75·5	0·554	0·408	0·92
9	0·78	76·0	0·557	0·386	0·91	0·88	78·0	0·568	0·468	1·30
12	0·82	77·7	0·566	0·411	1·11	0·96	79·0	0·574	0·538	1·63
15	0·84	78·5	0·571	0·419	1·22	1·00	81·0	0·584	0·552	1·95

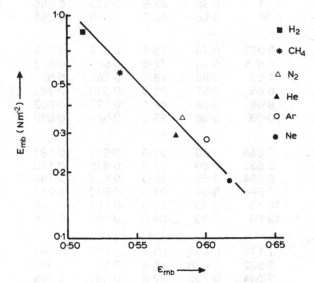

Fig. 7.4. Ln E_{mb} versus ϵ_{mb} for six gases.

7.5 THE EFFECT OF GRAVITY

A well-known method of proving a theory is to vary the relevant parameters independently and to verify whether the theory correctly predicts the experimental results. According to the theory

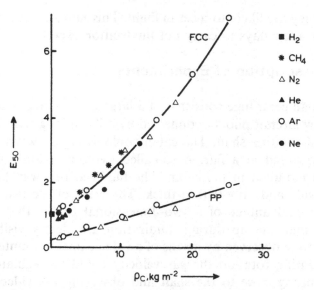

Fig. 7.5. E_{50} versus gas density ρ_c as found with fresh cracking catalyst (FCC) and polypropylene (PP).

presented in this and the foregoing chapter, the maximum stable bed porosity ε_{mb} is found from

$$N_F = \frac{(\rho_d - \rho_c)^2 \rho_d d_p^4 g^2}{\mu^2 E_{mb}} = \left\{ \frac{180(1 - \varepsilon_{mb})}{\varepsilon_{mb}^2 (3 - 2\varepsilon_{mb})} \right\}^2 \qquad (7.3)$$

in which E_{mb} is given by eqn (7.2).

For a particular powder the only parameters that can be varied independently are the gravity (or the apparent gravity) and the gas viscosity. Presumably, the largest variations in the maximum stable bed porosity ε_{mb} would be found if we could decrease the gravity. This, however, would mean going to the Moon where the gravity is one-sixth of that on Earth. As this seems hardly feasible for some time to come, we have to resort to increasing the gravity by centrifugal methods. In the National Aerospace Medical Centre in Soesterberg, in the Netherlands, a so-called human centrifuge is available for the testing and training of military aircraft pilots. In this equipment they can be exposed to the increased values of gravity they are likely to meet in flight. This human centrifuge was hired for a few days to carry out fluidization experiments.

7.5.1 Description of Experiments

The human centrifuge consisted of a large cabin—large enough to seat a few aircraft pilots—connected by a 2·5 m long radial arm to a vertical rotating shaft. The cabin could swing outwards as in a merry-go-round at a fair. A so-called two-dimensional fluidized bed was mounted in the cabin. The fluidized bed was 1 m high, 0·25 m wide and only 0·01 m thick. The walls of the bed were of glass. The advantage of a two-dimensional bed is that the gas bubbles that rise up during fluidization are clearly visible. The bubbles were observed by means of a video camera mounted in the cabin. During rotation the gas velocity could be regulated by a person sitting close to the shaft and observing the video screen. The apparent gravity g_a can be calculated from

$$g_a = (g^2 + \omega^4 r^2)^{0.5}$$

In this equation ω is the angular velocity and r the radius at which the fluidized bed is rotating. The gravity g_a was also measured by means of a specially designed apparent-gravity meter (for further details see Mutsers & Rietema, 1977). Since, during rotation, g_a varies over the height of the bed, only an average bed porosity is measured, whereas instability of the bed first arises at the top of the bed where the porosity is highest (see Fig. 7.6). The measured bed porosity must be corrected for this variation.

Two powders were investigated in the human centrifuge:

(1) a non-porous polypropylene powder with a particle density of 920 kg m^{-3} and an average particle size of 40 μm;
(2) a spent cracking catalyst powder with a particle density of 1414 kg m^{-3} and an average particle size of 62 μm.

These powders were both fluidized with nitrogen as well as with hydrogen. The results are plotted in Fig 7.7 and 7.8 as ε_{mb} versus the ratio of g_a over g. Also the theoretical lines according to eqns (7.2) and (7.3) (solid curves) are indicated. There is good agreement with the experiments.

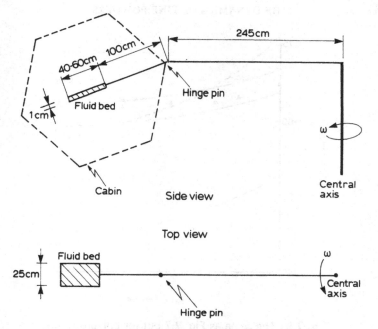

Fig. 7.6. Schematic diagram of human centrifuge.

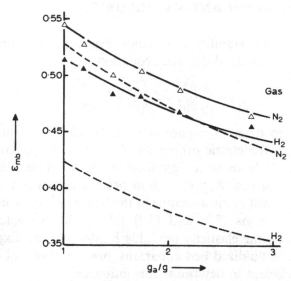

Fig. 7.7. ϵ_{mb} versus variation of apparent gravitational constant as measured for spent cracking catalyst fluidized in a centrifugal field with nitrogen ($\triangle$) and hydrogen ($\blacktriangle$) respectively: ——, theoretical line Rietema/Mutsers; ----, theoretical line Foscolo/Gibilaro.

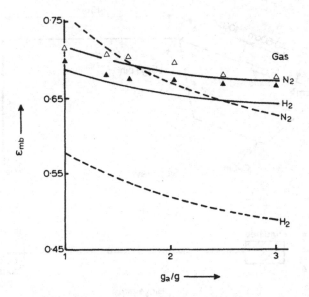

Fig. 7.8. The same as Fig. 7.7 but for polypropylene.

7.6 DISCUSSION AND CONCLUSION

The theory of the stability of a homogeneous gas-fluidized powder bed which led to the differential equation

$$\frac{\partial^2 \varepsilon}{\partial t^2} - \frac{E}{\rho_d}\frac{\partial^2 \varepsilon}{\partial h^2} + C\left(\frac{\partial \varepsilon}{\partial t} + U_c\frac{\partial \varepsilon}{\partial h}\right) = 0 \qquad (6.23)$$

enables us to derive the parameter E, which is the quantitative expression of the elastic properties of the powder packing during fluidization. It leads to a significant relation between E and the porosity of the bed. Any change in process conditions of the gas fluidized bed will cause a change in the bed porosity, which can be predicted by eqns (7.2) and (7.3) taking the influence of gas adsorption on the elasticity modulus E into account. Experiments with a simple fluidized bed apparatus, one at 1 bar and one at 10 bar, are sufficient to determine this influence.

In eqn (6.23) the minus sign before the second term is crucial. If it were positive, the bed would appear to be unstable at all porosities above the packed bed porosity. Hence it can be concluded that elasticity of the powder packing is a necessary condition for the bed to be able to expand homogeneously at a not too high porosity.

Several investigators (Verloop (1970), Oltrogge (1972) and Foscolo & Gibilaro (1984, 1986)) have tried to explain the necessary elasticity of the bed by means of hydrodynamics. When their theories are carefully worked out in the form of a differential equation such as eqn (6.23), in all these theories the term E appears with a positive sign. This would imply instability at all porosities.

True elastic deformation of a system is always completely reversible and no energy is dissipated. In hydrodynamics, however, flow and movement are always accompanied by energy dissipation. Therefore, it can be concluded that elastic forces based on hydrodynamics cannot exist. Accordingly, any stability theory based solely on hydrodynamics is doomed to fail.

In accordance with the theory on interparticle forces developed in Chapter 4, it is suggested that the effect of pressure on the stability of homogeneous gas-fluidized beds of fine powders is caused by gas adsorption to the surface of the solid particles. This gas adsorption increases the interparticle forces and, hence, also the elasticity modulus E. In Chapter 8 a model will be presented which describes quantitatively the relation between the elasticity and interparticle forces. This model describes also the dependence of the elasticity modulus on the porosity.

REFERENCES AND BIBLIOGRAPHY

Foscolo, P.U. & Gibilaro, L.G. (1984). A fully predictive criterion for the transition between particulate and aggregative fluidization. *Chem. Eng. Sci.*, **39**, 1667.

Foscolo, P.U. & Gibilaro, L.G. (1986). The influence of gravity on the stability of fluidized beds. *Chem. Eng. Sci.*, **41**, 2438.

Mutsers, S.M.P. & Rietema, K. (1977). Gas–solids fluidization in a centrifugal field. The effect of gravity upon bed expansion. *Powder Techn.*, **18**, 249.

Mutsers, S.M.P. & Rietema, K. (1979). Direct measurement of the mechanical response of a homogeneously fluidized bed to vertical vibration of an immersed object. *Powder Techn.*, **24**, 57, 65.

Oltrogge, R.D. (1972). *Gas fluidized bed of fine particles*. PhD Thesis, University of Michigan.

Piepers, H.W., Cottaar, E.J.E., Verkooyen, A.H.M. & Rietema, K. (1984). Effects of pressure and type of gas on particle–particle interaction and the consequences for gas–solid fluidization behaviour. *Powder Techn.*, **37**, 55.

Rietema, K. & Piepers, H.W. (1990). The effect of interparticle forces on the stability of gas-fluidized beds. I. Experimental evidence. *Chem. Eng. Sci.*, **45**, 1627.

Verloop, J. & Heertjes, P.M. (1970). Shock waves as a criterion for the transition from homogeneous to heterogeneous fluidization. *Chem. Eng. Sci.*, **25**, 825.

8

A Model of Expanded Powders

NOTATION

A Hamaker constant (J)

d_p Particle diameter (m)

E Elasticity modulus (N m^{-2})

F_y Elastic force at asperity contact (N)

F_λ Force at particle contact (N)

k Coordination number (—)

n Number of particles per unit volume of powder packing (m^{-3})

N Number of particles in volume V (—)

N_c Number of particle contacts in volume V (—)

R_a Radius of an asperity (m)

V_p Volume of a particle (m^3)

V_ε Total pore volume in volume V (m^3)

γ Proportionality constant (—)

ε Porosity of packing (—)

λ Thickness of a layer of particles (m)

σ_e Stress operating on the packing (N m^{-2})

8.1 INTRODUCTION

On the basis of the theory developed in Chapters 4 and 6 and the experimental results discussed in Chapter 7, we now arrive at a model of homogeneous fluidized powders which will be elaborated in this chapter.

When the gas velocity through a homogeneous fluidized bed is increased the powder bed expands as long as the gas velocity is below the critical velocity at the bubble point. The expansion of the bed, however, does not occur continuously but happens in small steps. During each step the elasticity modulus is constant, but when the expansion becomes too large due to a further increase of the gas velocity the elasticity limit of the powder bed is exceeded and some contact points between the powder particles will be broken. The powder bed then seeks a new structure with a somewhat higher porosity, with less contact points and with a new but lower elasticity modulus.

On the other hand when the gas velocity is decreased the drag force on the particles is reduced and the equilibrium between drag force and weight of the particles is broken. As long as the deviation of the equilibrium is not too large it can be compensated by the elastic force. When, however, the deviation becomes too large the elasticity limit is exceeded and the powder structure seeks to rearrange itself and forms new contact points between the particles, now at a lower porosity. This is attended by an increase of the elasticity modulus.

From the above it follows that there is a direct relation between the elasticity modulus and the coordination number. In the following this concept will be further elaborated.

8.2 THE RELATION BETWEEN ELASTICITY AND POROSITY

The powder elasticity is defined as the ratio between the stress and the porosity change caused by this stress:

$$\sigma_e = E \cdot \Delta \varepsilon \qquad (8.1)$$

We shall consider a cube of powder of unit volume containing n layers of n^2 particles with volume V_p. The porosity of the powder is ε. An outward force will stretch the cube of powder in the direction of this force with the consequence that the porosity increases with $\Delta\varepsilon$. The thickness of each layer must increase with $\Delta\lambda$ where λ is the thickness of an undisturbed layer. The relation between $\Delta\varepsilon$ and $\Delta\lambda$ follows from

$$(1-\varepsilon)=(1-\varepsilon-\Delta\varepsilon)(1+n\Delta\lambda)$$

Neglecting $n \cdot \Delta\lambda \cdot \Delta\varepsilon$ we find that

$$\Delta\varepsilon=(1-\varepsilon)\frac{\Delta\lambda}{\lambda}=(1-\varepsilon)n\Delta\lambda \qquad (8.2)$$

Suppose now that this porosity increase is achieved by a force F_λ operating at each contact point. It then follows that

$$\sigma_e=n^2F_\lambda\cdot\frac{k}{2} \qquad (8.3)$$

in which k is the coordination number. Since

$$n=\frac{1}{\lambda}=\frac{1}{d_p}\sqrt[3]{\frac{6(1-\varepsilon)}{\pi}}$$

it can finally be derived that

$$E=\frac{F_\lambda}{\Delta\lambda}\frac{k}{2d_p}\sqrt[3]{\frac{6}{\pi(1-\varepsilon)^2}} \qquad (8.4)$$

When for a certain powder the stress is not too high, the ratio $F_\lambda/\Delta\lambda$ will be constant and independent of the porosity. Hence the porosity dependence of the elasticity is given by

$$E/E_0=k\sqrt[3]{\frac{1}{(1-\varepsilon)^2}} \qquad (8.5)$$

8.3 THE COORDINATION NUMBER

It was suggested by Smith (see Manegold, 1955) that the relation between k and the porosity is given by $k\varepsilon = 3 \cdot 1$. Indeed for the rhombohedral packing $k\varepsilon = 3 \cdot 114$ and for the cubic packing $k\varepsilon = 2 \cdot 858$. It will be clear, however, that such a relationship cannot be true for $\varepsilon \to 1$ since in that case k must go to zero. Besides, in practice $k = 2$ must be an absolute minimum since in that case the particles are stacked on top of each other without any sideward support and so have lost their coherence. Hence, we introduce $k^* = k - 2$.

We shall now consider a cube of N particles with volume V_p. The total number of contact points $N_c = Nk/2$. The total pore volume of the cube is $V_\varepsilon = NV_p\varepsilon/(1-\varepsilon)$. Suppose that one contact point is broken as result of the stress tensor operating on the cube of powder. The total pore volume will increase by ΔV_ε. The total volume of the powder will also increase by this amount, while the porosity is increased by $d\varepsilon$.

It can be expected that ΔV_ε is related directly to the volume of one particle and is inversely proportional to $(k-2)$. Hence we propose that

$$\Delta V_\varepsilon = \gamma \frac{V_p}{k-2} \tag{8.6}$$

in which γ is a numerical constant. When $-dN_c$ contacts are broken (dN_c is negative!) the increase of the total pore volume must be equal to

$$-(\Delta V_\varepsilon) \cdot dN_c = \frac{d}{d\varepsilon}(V_\varepsilon)\, d\varepsilon = \frac{NV_p}{(1-\varepsilon)^2}\, d\varepsilon$$

After substitution of eqn (8.6) and of $dN_c = N\,dk/2$ it follows that

$$-\gamma \frac{dk}{2(k-2)} = \frac{d\varepsilon}{(1-\varepsilon)^2} \tag{8.7}$$

After integration the result is

$$-\frac{\gamma}{2}\ln(k-2) = \frac{1}{1-\varepsilon} + C \tag{8.8}$$

We now return to eqn (8.5) but here too we replace k by k^*. This is justified by the following reasoning. Consider three pairs of contacting particles as they occur in a random packing. The three configurations **a**, **b** and **c** (see Fig. 8.1) will all be subject to the

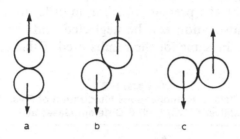

Fig. 8.1. Three configurations of two neighbouring particles.

same vertical stress. In configuration **a** the increase in the vertical distance L caused by this stress cannot be more than the maximum increase $r_{ss} \cdot \Delta \omega$ (see Chapter 4) as long as the contact between the particles is not broken. It has, therefore, practically no effect on the relation between F_λ and $\Delta \lambda$. By replacing k by k^* these contacts are excluded from the further derivation (the foregoing means that also in eqn (8.4) k should be replaced by k^*). We now rewrite eqn (8.5):

$$E/E_0 = k^* \sqrt[3]{\frac{1}{(1-\varepsilon)^2}} \tag{8.5a}$$

We apply it for two different values of E as found for a specific powder by fluidization with two different gases:

$$k_1^*/k_2^* = \frac{E_1}{E_2} \sqrt[3]{\left(\frac{1-\varepsilon_1}{1-\varepsilon_2}\right)^2} \tag{8.9}$$

From this equation the ratio k_1^*/k_2^* can be determined. Applying eqn (8.8) twice yields

$$\gamma = 2 \left\{ \frac{1}{(1-\varepsilon_1)} - \frac{1}{(1-\varepsilon_2)} \right\} \frac{1}{\ln(k_1^*/k_2^*)} \tag{8.10}$$

Hence, also γ can be determined. The constant C must be determined by trial and error, e.g. by assuming $k^* = 6$ at $\varepsilon = 0.40$.

The foregoing analysis was applied to the results obtained with fresh cracking catalyst presented in Chapter 7. These results are summarized again in Table 8.1 but only for those experiments that were carried out at a pressure of 1 bar, in order to be sure that the effect of gas adsorption can be neglected and that the reference elasticity E_0 is the same for the results used. Table 8.2 presents the

TABLE 8.1
Relevant Data of Homogeneous Fluidization of Fresh Cracking Catalyst (FCC) with 6 Different Gases at 1 bar

Type of gas	ε_{mb}	u_{mb} $(cm\,s^{-1})$	E_{mb} $(N\,m^{-2})$
Hydrogen	0.511	0.80	0.850
Helium	0.578	0.57	0.293
Methane	0.538	0.70	0.560
Neon	0.617	0.50	0.182
Nitrogen	0.583	0.632	0.351
Argon	0.600	0.590	0.278

TABLE 8.2
The Ratio k_1^*/k_2^* and the Value of γ as Derived for 6 Gas combinations

Gas combination	k_1^*/k_2^*	γ
Argon-methane	0.451	0.842
Neon-hydrogen	0.1819	0.664
Nitrogen-methane	0.585	0.854
Neon-nitrogen	0.490	0.596
Argon-nitrogen	0.286	0.728
Helium-hydrogen	0.298	0.536

values of k_1^*/k_2^* and γ as derived for six different combinations of these experiments. The average value of γ is determind at 0.704. With this average value of γ the theoretical dependence of E on ε is plotted in Fig. 8.2, assuming $k^* = 6$ at $\varepsilon = 0.40$, which yields for the constant C a value of -2.297. Also the empirical curve $E/E_0 = \exp[15(0.50 - \varepsilon)]$ as well as the experimental results are

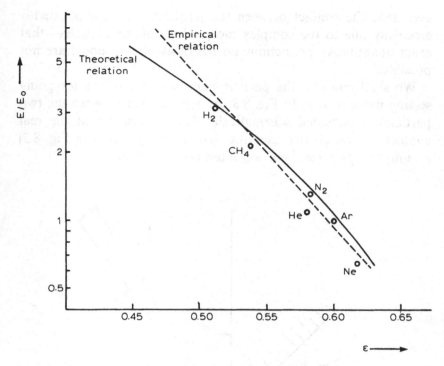

Fig. 8.2. E/E_0 versus the porosity; empirical and theoretical relations.

given in this figure. The figure shows that the strong dependence of the elasticity modulus E on the porosity can be satisfactorily explained by assuming that the origin of the powder elasticity is to be found in the existence of Van der Waals forces between the particles of the powder bed.

8.4 THE RELATION BETWEEN COHESION AND ELASTICITY

A basic assumption in all of the preceding considerations has been that there must be a relation between cohesion and elasticity as both are determined by the interparticle forces. In this section a model of this relation will be presented. It must be stressed, however, that the contact between the particles is so complicated—especially due to the complex morphology of the particles—that exact quantitative predictions on the basis of this model are not possible.

We shall consider the particle pairs **b** and **c** of the foregoing section more closely. In Fig. 8.3 the contact area between the two particles is indicated schematically. It is assumed that the real contact is through two asperities (shown exaggerated in Fig. 8.3) although in practice there will often be more than two.

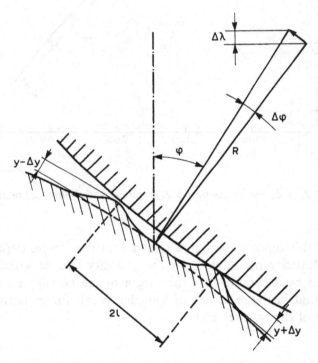

Fig. 8.3. Schematic diagram of contact between two particles via asperities.

The distance between the asperities is put equal to $2l$. As a result of the external force one of these asperities will be lengthened by Δy while the other will be compressed by the same amount. Following the theory developed in Chapter 4, y is put equal to

$z - h = r_{ss}\omega$. Owing to this mechanism the contact angle between the two particles will change over $\Delta\varphi$ and the vertical distance over $\Delta\lambda$:

$$\Delta\lambda = R \cdot \Delta\varphi \cdot \sin\,\varphi$$

with $\Delta\varphi = \Delta y/l$. Averaged over all possible configurations this gives

$$\overline{\Delta\lambda} = \frac{\pi R}{4} \cdot \frac{\Delta y}{l} \tag{8.11}$$

The moment M exerted on the particles by the external force is $F_\lambda d_p \sin\,\varphi$; the averaged $\overline{M} = \pi F_\lambda d_p/4$. The moment exerted by the two asperities is $2F_y l$. These two moments must be equal. This yields

$$F_\lambda = \frac{8}{\pi}\frac{l}{dp}F_y \tag{8.12}$$

Substitution of eqns (8.11) and (8.12) in eqn (8.4) gives

$$E = \frac{32}{\pi^2}k^* \frac{l^2}{d_p^3}\left(\frac{F_y}{\Delta y}\right)\sqrt[3]{\frac{6}{\pi(1-\varepsilon)^2}} \tag{8.13}$$

F_y can be calculated from eqn (4.7a) in Chapter 4 with a value $\phi = 1.85$, which holds for the maximum external force without breaking the contact.

$$\phi = \frac{6r_{ss}^2 F_y}{A R_a}$$

In this equation R_a is the radius of an asperity.

When for the relevant powder the parameter $\psi = 10$, the decrease in the flattening of the asperities amounts to $\Delta y = 0.2\,r_{ss}$. Hence

$$F_y/\Delta y = 1.54\frac{A R_a}{r_{ss}^3}$$

When we assume a porosity of 0.50, then from Table 8.2 it follows that $k^* = 2.325$. With $d_p = 60\,\mu m$, $A = 10^{-19}\,J$ and $r_{ss} = 3.5 \times 10^{-10}\,m$

it is found that

$$E = 0.25 \, (l^2 R_a) \times 10^{24} \, \text{N m}^{-2}$$

It will be clear that the result found for E is strongly determined by the choice made for l and R_a. Unfortunately, there is not enough information available on the distribution of the asperities over the surface of the particles and this information will be difficult to obtain.

With a choice of $l = 200$ Å (1 Å $= 10^{-10}$ m) and $R_a = 100$ Å, it is found that $E = 1.0 \, \text{N m}^{-2}$ which at least is of the right order of magnitude. It is believed, therefore, that on the basis of the model presented it is possible to explain the values of E as derived from the experiments discussed in the foregoing chapter.

Nevertheless, any results for E on the basis of assumptions about l and R_a are as yet purely speculative.

According to Rumpf (1958) (see also Molerus, 1975), the tension cut-off σ_c is given by

$$\sigma_c = \frac{kF_y}{\pi d_p^2} (1 - \varepsilon) \tag{8.14}$$

It must be remarked that in this equation k should not be replaced by k^* as particles in configuration **a** (see Fig. 8.1) certainly make a strong contribution to the tension cut-off. For the ratio E/σ_c it is now found that

$$E/\sigma_c = \frac{32}{\pi} \left(\frac{k-2}{k} \right) \left(\frac{l^2}{\Delta y \cdot dp} \right) \sqrt[3]{\frac{6}{\pi} \frac{1}{(1-\varepsilon)^5}}$$

$$= 12.6 \left(\frac{k-2}{k} \right) \left(\frac{l^2}{\Delta y \cdot dp} \right) \sqrt[3]{\frac{1}{(1-\varepsilon)^5}} \tag{8.15}$$

With $l = 200$ Å, $\Delta y = 1$ Å and $d_p = 50 \, \mu\text{m}$ as an example, the dependence of E/σ_c on the porosity is shown in Table 8.3. A notable point in this table is that at not too high a porosity the ratio E/σ_c does not differ much from 1.

Another conclusion that can be drawn from eqn (8.15) corresponds well with the remark made in Section 6.6 on the phenomenon of channelling. It was suggested there that it is in particular

TABLE 8.3

The Theoretical Coordination Number k and the Ratio of the Elasticity Modulus E to the Cohesion Constant σ_c as a Function of the Porosity ϵ

ϵ	k	E/σ_c
0·40	8	1·77
0·45	5·897	1·80
0·50	4·325	1·72
0·55	3·237	1·46
0·60	2·577	1·04
0·65	2·204	0·54

the ratio E/σ_c that determines whether a powder will show channelling during fluidization or not. In Chapter 4 it was shown that when a powder is 'stretched' the increase of the distance Δy between neighbouring particles is larger when the powder is less hard, that is to say that the Young's modulus of the solid has a lower value although the cohesion force is independent of the hardness.

Equation (8.15) indeed shows that the ratio E/σ_c decreases when Δy increases. At present, however, too little information on channelling is available to confirm this theory. Also no experimental technique is known for measuring Δy directly.

REFERENCES

Manegold, E. (1955). *Kapillar Systeme*, Vol. 1., Strassenbahn Chemie und Technik, Heidelberg, Germany.

Molerus, O. (1975). Theory of yield of cohesive powders. *Powder Techn.*, **12**, 259.

Rumpf, H. (1958). Grundlagen und Methoden des Granulierens. *Chemie-Ing. Techn.*, **30**, 144.

9

Rheology of Fluidized Powders†

NOTATION

d_p	Particle diameter (—)
G	See eqn (9.1) (s^{-1})
M_d	Mass rate of powder $(kg\ s^{-1})$
p	Pressure $(N\ m^{-2})$
Q_d	Volumetric powder flow rate $(m^3\ s^{-1})$
Q_g	Gas flow rate through bypass $(m^3\ s^{-1})$
r	Radial coordinate (m)
R	Radius of standpipe (m)
u_{ao}	Fluidization velocity in vessel A $(m\ s^{-1})$
u_{bo}	Fluidization velocity in vessel B $(m\ s^{-1})$
v_d	Average velocity of powder $(m\ s^{-1})$
v_s	Slip velocity $(m\ s^{-1})$
v_w	Velocity of particles at the wall $(m\ s^{-1})$
z	Vertical coordinate (m)
ε	Porosity (—)
μ	Gas viscosity $(N\ s\ m^{-2})$

†Substantial parts of this chapter were taken from Dr G. van den Langenberg-Schenk's doctoral thesis, with her permission.

162

μ_w Kind of wall friction viscosity $(N\,s\,m^{-3})$
ρ_d Density of solid particles $(kg\,m^{-3})$
τ_w Shear stress at the wall $(N\,m^{-2})$
τ_{w0} Wall yield stress $(N\,m^{-2})$
τ_0 Yield stress of powder $(N\,m^{-2})$

9.1 INTRODUCTION

Although many researchers have investigated the rheology of gas-fluidized powders no reliable theoretical description is as yet available. Various measuring methods have been applied, varying from Stormer, Brookfield or Couette viscometers, torsion pendulum or floating/falling ball-methods to the study of average flow velocity through open inclined or horizontal channels.

The frequent use of conventional viscometers is readily understandable since in a fluidized bed a resistance against flow exists, just as in liquids. This resistance is a kind of internal friction between the particles and resembles the concept of 'viscosity' used in describing the rheology of liquids. The friction is caused by static forces such as electrostatic forces, Van der Waals forces and capillary forces, as well as by dynamic forces which are due to the particle motion. It will be clear that the shape of the particles, the interparticle distances (which are directly related to the bulk porosity), the gas velocity and the gas viscosity will influence the 'powder viscosity'.

When the fluidized bed is freely bubbling additional problems occur. In that case the fluidized bed consists of two distinct phases: a dense phase and a bubble phase. Because there are no particles in the bubble phase and the momentum of a bubble can be neglected, it makes no sense to ascribe a viscosity to the bubble phase. Hence, in the following we mean by 'viscosity' the viscosity of the dense phase.

Nevertheless, the presence of the bubbles creates a series of problems that complicate the measurement of this viscosity:

(1) overall circulation of the dense phase will be induced by the bubbles (see Chapter 10);

(2) when a bubble hits the measuring device the shear stress will locally fall to zero;

(3) rising bubbles create extra momentum transport causing boundary layers to break up near the measuring device;

(4) depending on the measuring method the bubbles might influence the motion of the measuring device;

(5) the conditions for which the usual relationship between shear stress and shear rate holds (i.e. laminar flow) are strongly disturbed.

For all of these reasons measuring the viscosity in a bubbling fluidized bed is pointless. It is a great pity that most authors do not recognize this problem and do not mention whether the bed in which they carried out measurements was freely bubbling or not. Only from the gas flow rate can one guess that in many cases the bed was indeed freely bubbling.

All methods described in the literature that use an extra measuring device rely on the assumption that the powder sticks to the surface of that device. However, when this surface is smooth and hard it is questionable whether this is correct and whether slip of the powder will not occur there, i.e. whether the powder layer adjacent to the surface will not have a velocity differing from that of this surface. It should always be verified whether slip occurs or not but most authors ignore this phenomenon. Another shortcoming sometimes encountered is that assumptions are made concerning the rheological behaviour, e.g. Newtonian behaviour, or concerning the flow regime e.g. laminar flow. Often investigations are reported that do not allow the validity of these assumptions to be checked. The evaluation of the measurement results, therefore is sometimes doubtful.

The equations of Grace (1970) are based on measurements of the shape of bubbles rising in a fluidized bed. He compares this shape with that of gas bubbles rising in liquids and relates it to the Reynolds number of the rising bubble from which the powder viscosity can be derived. This seems a very attractive method since no disturbing devices are inserted in the bed. The pressure field around a gas bubble in a fluidized bed, however, is not comparable to that around a gas bubble in liquids. This is caused by the flow of gas (from the dense phase) through the gas bubble. This effect will certainly influence severely both the rising velocity and the shape of the bubble.

9.2 LITERATURE SURVEY

In Tables 9.1 and 9.2 the details of investigations reported in the literature are summarized. From these investigations some general trends can be derived:

(1) the viscosity or apparent viscosity is of the order of magnitude of 0·1 to 10 poise;
(2) the viscosity decreases with increasing gas flow rate, i.e. with increasing porosity;
(3) the viscosity decreases with decreasing particle size, at least in the particle range that has been investigated by the various authors.

From the foregoing it will be clear that further investigation of the rheology of fluidized powders is justified and that in such an investigation the phenomenon of bubbling should be avoided.

In the following a study of the rheology of homogeneously fluidized solids is described. It was carried out in a vertical cylindrical standpipe of glass in which the slip velocity of the solids along the pipe wall could be measured from the outside. The wall shear stress, the bulk velocity and the average porosity could be measured and these quantities could be correlated by means of a modified Rabinowitz equation.

9.3 DESCRIPTION OF THE APPARATUS

Essentially, the equipment consists of two cylindrical fluidized beds, one above the other, connected by an exchangeable vertical standpipe of glass coaxial with the fluidized beds (see Fig. 9.1). The length of the standpipe between the two vessels was 142 cm, its internal diameter 6·0 cm. In the lower vessel the standpipe extended into the lower bed with a brass dipleg, the total length of

Table 9.1

Measurements with Conventional Methods in Stationary, Air Fluidized Beds

References	Bed diameter (cm)	Measuring method	Powder	$\bar{d}_p$ (µm)	Size range
Ashwin et al. (1960)	6·6	torsion	graphite coated	125·4/ 304·8	narrow
Hagyard & Sacerdote (1966)	7·6	pendulum	shellac spheres		
Grace (1970)	14	rising bubble	ballotini	60/550	narrow
			silversand	72/500	narrow
			synclyst catalyst	52	wide
			magnesite	240	wide
Leont'ev & Vakhrushev (1976)	10·5	floating ball	quartz sand	220	narrow
Matheson et al. (1949)	4·6	Stormer	synthetic cracking catalyst (spherical)	47/254	narrow
			catalyst (irregular)	45/456	narrow
			sand	28/96	narrow
			metal oxide (spherical)	163	narrow
			iron (irregular)	33	narrow
Kramers (1951)	8·6	Stormer	riversand	≈130	rather wide
				≈180	rather wide
Omae & Furukawa (1954) Furukawa & Omae (1958)	6	Stormer	polyvinyl acetate beads	277/755	narrow
Diekman & Forsythe (1953)	10	Brookfield	cracking catalyst	57/73	narrow

Table 9.1—*contd.*

Fluidization	Remarks	Viscosity (range) (Poise[a])
heterogeneous	$1 < u_0/u_{mf} < 12$; Viscosity decreases with increasing u_0 and with decreasing $\bar{d}_p$	0·5–3
heterogeneous	Viscosity of dense phase is relevant quantity. Viscosity decreases with decreasing $\bar{d}_p$	7–9·5 8–12 4 9
homogeneous + heterogeneous	Theoretical expression shows decrease in viscosity with increasing porosity	1·8–5·4
heterogeneous	Viscosity decreases as superficial gas velocity increases and particle size and particle density decrease; spherical particles show larger viscosity than irregularly shaped particles; paddle-speed 200 rpm	0·16–1·6 0·06–3·4 0·8–2·1 3·2 2·6
heterogeneous	$1·7 < u_0/u_{mf} < 2·9$ Paddles replaced by a dumbbell; speed < 30 rpm. Viscosity decreases with increasing u_0/u_{mf}	16–64 18–48
homogeneous + heterogeneous	Paddle speed 113 rpm; viscosity decreases as flow rate increases and $\bar{d}_p$ decreases	0·7–22·5
heterogeneous	Channelling and bubbling occurred in the bed. Higher particle density and coarser particles cause higher viscosity	0·26–6·5

Table 9.1—*contd.*

References	Bed diameter (cm)	Measuring method	Powder	$\bar{d}_p$ (µm)	Size range
Fa-Ke Liu & Orr (1960)	4·45	Brookfield	glass beads	44/123	narrow
			SiAl cracking catalyst	45	narrow
			polystyrene beads	349	narrow
Van der Leeden & Bouwhuis (1960)	1·0/ 5·0	Brookfield	spent cracking catalyst	≈70	wide
Botterill et al. (1972)	14	Brookfield	bauxilite	102	narrow
			silica sand	80/300	wide
			zircon sand	≈150	wide
Woodruff (1973)	6	Brookfield	silica powder	20/60	narrow
			titanium powder	25/50	narrow
Schügerl (1973)	7·5 13·25	Couette	glass beads	50/500	narrow
			alumina plates	90	—
Schügerl et al. (1961)	14·5		cork	≈50	wide
			polystyrol	250	narrow
			quartz	75/450	narrow
			silicon carbide	45/190	narrow
Lehmann et al. (1973)	19 36	Couette	quartz	160	narrow
Ritzmann & Schügerl (1974)	50				

Table 9.1—*contd.*

Fluidization	Remarks	Viscosity (range) (Poise[a])
heterogeneous	Two concentric, thin walled brass cylinders were rotated. Viscosity decreases with increasing flow rate, decreasing particle size, decreasing bed weight	1–4·6 0·3–1·8 1–4·6
particulate	Rotation speed up to 100 rpm; torque independent of rotation speed (0·5–50 rpm); independent of roughness of cylinder, of bed diameter (6·5–9·0 cm), does not change when water is substituted for air!	No values for μ were given, only the torque in arbitrary units
heterogeneous	A hollow cylinder is used	No absolute values given 15–80 arbit.units 40–70 arbit.units 0·16–3·75 0·16–0·2
packed bed/ homogeneous heterogeneous		1·4–67 — — 0·9–2·5 5·8–261 4·8–34
heterogeneous	Measured viscosity depends on location in the bed	

[a]1 Poise = 10^{-1} Nsm^{-2}.

Table 9.2

Measurements of Air Fluidized Powders Flowing through Open Inclined or Horizontal Channels

References	Channel length (m)	Channel width (mm)	Powder	$\bar{d}_p$ (µm)	Size range
Mori et al. (1955)	85	50	sand	200	narrow
			alumina	37	wide
			bauxite	90	wide
Siemens & Hellmer (1962)	2·0	150	quartz sand	217	wide
Turcajova & Neuzil (1976, 1977)	0·808 1·2	43	corundum glass beads	500 1000	narrow narrow
Neuzil & Turcajova (1969, 1977)			sand	1100	narrow
Botterill & van der Kolk (1971)		104/288	dune sand bauxilite	138/185 102	wide narrow
Botterill & Bessant (1973, 1975)			catalyst ash	77 380/590	narrow wide
Bessant & Botterill (1973)					
Botterill & Abdul Halim (1978, 1979)					
Muskett et al. (1973)	2·4	75	sand	150	—
McGuigan & Pugh (1976)	3·0	100/150	sand	150	wide

Table 9.2—*contd.*

Channel slope (degree)	Fluidization range u_0/u_{mf}	Viscosity (Poise[a])	Remarks
1–15	—	—	Only superficial gas velocities are given: sand: 5·6–11·3 cm s^{-1} alumina: 1·0–2·6 cm s^{-1} bauxite: 3–7 cm s^{-1}
1–6	3–6	1–9	
1–7	1·3–2·5	—	Only relative viscosities are given: $\mu(u_0)-\mu(2\cdot5\,u_0)$ value ranges from 1 to 3·5
0	2–4 2–3·5 1·75–3 1·5–2·25	 ≈1 ≈1	A closed horizontal channel is used, flow is caused by paddles. Viscosity values only order of magnitude.
4–8	—	—	Only the superficial gas velocity is mentioned, 1·5–4·15 cm s^{-1}
0–30	1·6–5·5	1–1·5	

Table 9.2—*contd.*

References	Channel length (m)	Channel width (m)	Powder	$\bar{d}_p$ (µm)	Size range
Shinohara et al. (1974)	1·2	42	glass beads	60–80 mesh	—
Woodcock & Mason (1976)	6	100	Corvic	140	wide
Singh et al. (1978)	0·75 1·5	41	sand	241	wide
Ishida et al. (1980)	0·954	39	glass beads	160/390	narrow/ wide
			porous alumina	230	narrow
			sand	190	narrow

standpipe and dipleg being 220 cm. Without the dipleg no stable powder flow was possible. There is the additional advantage that, after each experiment, the fluidized solids could be transported back into the upper vessel by reversing the gas pressure difference between the two vessels. Stable powder flow appeared to be possible only during downward flow of the powder. The fluidized beds were contained in two stainless steel vessels each of about 200 litre volume and with an external diameter of 60 cm. The gas distributors consisted of two perforated perspex plates and a porous flexolite plate (10 mm thick). This combination gave a homogeneous gas distribution and a sufficiently high pressure drop over the bottom plates. Each fluidized bed had its own closed gas circuit in which gas is circulated by means of a compressor.

When the fluidized powder flows downward, gas is entrained by the powder from the upper vessel towards the lower vessel. During an experiment the gas pressure in the bottom vessel would therefore increase at the expense of the pressure in the upper vessel, and the change in pressure drop across the standpipe would induce a change in the mass flow of the powder. In order to maintain a constant powder flow, a gas bypass has been provided from the

Table 9.2—*contd.*

Channel slope (degree)	Fluidization range u_0/u_{mf}	Viscosity (Poise[a])	Remarks
7·5–15		0·01–0·2	Only the superficial gas velocity is mentioned, 8–18 cm s^{-1}
0–12	2–12	—	
0–5	1·25–5	59	
14–23	0–3·7	—	

[a]1 Poise $= 10^{-1}$ Ns m^{-2}.

pressure side of the lower compressor to the suction side of the upper compressor. A needle valve controls the gas flow. The gas flow in the bypass determines the powder velocity in the standpipe. By means of a second bypass the pressure difference between the two vessels could be reversed in order to transport the powder from the lower to the upper vessel.

The standpipe could be more or less closed at both ends by means of two multi-blade butterfly valves (see Fig. 9.2). These valves were used to adjust the porosity in the downward flowing powder at a chosen average powder flow rate.

9.4 MEASUREMENTS

The bulk porosity ε of the powder flowing through the standpipe was measured by means of an electrical capacity probe, which consists of two flat brass plates (70 mm $\times$ 70 mm) mounted in a perspex housing, which can be moved up and down along the standpipe. The measured capacity of the system depends on the average bulk density of the flowing powder. The standard devia-

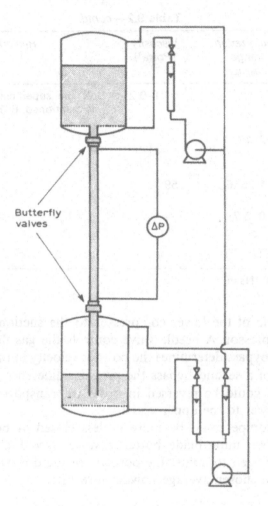

Fig. 9.1. Experimental setup.

tion in the measured porosity appeared to be less than 0·4%
Preliminary experiments with two capacity probes showed that
within the measuring accuracy no porosity gradient could be
detected over the length of the standpipe during an experiment.

The powder velocity at the wall (v_w) was measured by visual
observation of the flow of powder particles stained black. The

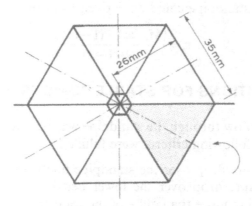

Fig. 9.2. Butterfly valve with six blades.

mass flow rate of the powder M_d was determined by measuring the increase of the pressure drop over the lower bed with time. This pressure drop increased linearly with time during more than 60% of the total transport time in one experiment, which indicates a constant powder flow rate.

The wall shear stress was derived from the pressure gradient in the standpipe (dp/dz). Preliminary experiments in which the pressure drop was measured over small sections of the standpipe indicated that the pressure gradient during transport was independent of height, which of course is in agreement with the observation mentioned above that there is no porosity gradient during transport. The wall shear stress τ_w can therefore be calculated from the formula

$$\tau_w = \frac{R}{2} \left\{ \frac{dp}{dz} - \rho_d (1 - \varepsilon) g \right\} \tag{9.1}$$

In this formula R is the internal radius of the standpipe while ρ_d is the density of the solids.

The gas flow Q_g through the bypass from the discharge side of compressor A to the suction side of compressor B was measured as well. With A the cross-sectional area of the standpipe, the average slip velocity v_s between the solids and the gas flowing through the standpipe could be derived from the formula

$$v_s = \frac{(M_d/\rho_d) - (1-\varepsilon)Q_g}{A\varepsilon(1-\varepsilon)} \tag{9.2}$$

9.5 CONDITIONS FOR STABLE POWDER FLOW

The powder flow through the standpipe was considered sufficiently stable if the following criteria were fulfilled:

(1) the pressure drop over the standpipe fluctuates less than 4%;
(2) the pressure drop over the lower bed increases linearly with time during more than 50% of the total transport time;
(3) within the measuring accuracy no porosity fluctuation is observed.

It will be clear that the three gas flows through the apparatus (the superficial fluidization gas flow u_{ao} in vessel A, the same (u_{bo}) in vessel B, and the gas flow through the bypass Q_g) have to be chosen carefully and adapted to each other in order to keep the powder transport under control. The combinations (u_{ao}^*, u_{bo}^*) that made stable homogeneous powder flow possible are indicated in Fig. 9.3. Four unsuitable combinations can be distinguished:

(1) $u_{bo} < u_{bo}^*$ and $u_{ao} > u_{ao}^*$: flow is very irregular, porosity decreases as well as M_d.
(2) $u_{bo} < u_{bo}^*$ and $u_{ao} < u_{ao}^*$: porosity in pipe becomes very low, transport stops.
(3) $u_{bo} > u_{bo}^*$ and $u_{ao} < u_{ao}^*$: as (2)
(4) $u_{bo} > u_{bo}^*$ and $u_{ao} > u_{ao}^*$: porosity increases, pressure drop fluctuates and bubbles appear in the pipe.

By varying the gas bypass flow and the upper butterfly valve opening, stable powder flow could be obtained at various porosities and at sufficiently low velocity to permit visual measurement of the wall slip velocity. The lower butterfly valve was completely open in all of these measurements to prevent an unknown amount of the powder in the standpipe from being supported by the valve and, thus, affecting the force balance over the powder.

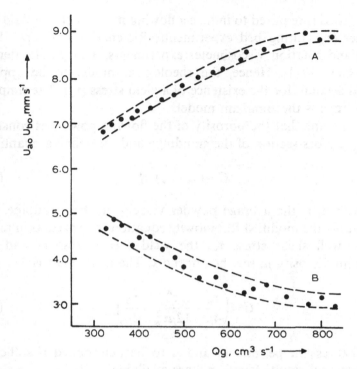

Fig. 9.3. Combinations of u_{a0} and u_{b0} in vessels A and B respectively that give stable flow.

About 50 stable-flow experiments at various bulk porosities were carried out with a fraction of fresh cracking catalyst ($\rho_d = 750$ kg m^{-3} and $d_p = 77$ µm). The experimental data are given in the thesis of Dr van den Langenberg-Schenk (1982) and also in her paper (1984) and will not be repeated here.

9.6 EVALUATION OF THE MEASUREMENTS

The measurements were evaluated by means of a modified Rabinowitz equation (see the appendix). The original Rabinowitz equation could not be used since it does not account for a possible slip velocity near the wall.

From other experiments it is known that a certain minimum

shear stress is required to induce a flowing movement in a fluidized powder (e.g. tilting bed experiments, Rietema & Mutsers, 1973, 1977; and rotation viscosimeter experiments, Van der Leeden & Bouwhuis, 1961). Hence, the rheological model to be applied should account for the existence of a yield stress (τ_0). The simplest model then is the Bingham model.

We assume that the porosity of the flowing powder is constant over the cross-section of the standpipe and we define a quantity

$$G = (\bar{v}_d - v_w)/R \tag{9.3}$$

in which $\bar{v}_d$ is the average powder velocity in the standpipe. We now apply the modified Rabinowitz equation to express G in terms of the wall shear stress τ_w, the yield shear stress τ_0 and the apparent viscosity μ; see the appendix. The final result is

$$G = \left(\frac{\tau_w}{4\mu} + \frac{\tau_0^4}{12\mu\tau_w^3} - \frac{\tau_0}{3\mu} \right)_\varepsilon \tag{9.4}$$

This expression permits τ_0 and μ to be determined if sufficient experimental results for (G, τ_w) are available.

With the assumption that τ_0 and μ can be related to the porosity by

$$\left. \begin{array}{l} \tau_0 = A + B\varepsilon \\ \mu = C + D\varepsilon \end{array} \right\} \tag{9.5}$$

the best regression is sought by means of the least-squares fit method. The final result is given by

$$\left. \begin{array}{l} -\tau_0 = 8\cdot10 - 15\cdot0\varepsilon \ \text{N m}^{-2} \\ \mu = 1\cdot59 - 2\cdot54\varepsilon \ \text{N s m}^{-2} \end{array} \right\} \tag{9.6}$$

while the square root of the average of $(G_{exp} - G_{calc})^2$ equals 0·0124.

In Figs 9.4 and 9.5 a comparison is made between the experimental results and the relation calculated using eqns (9.4) and (9.6). It is seen that the correlation is very good indeed.

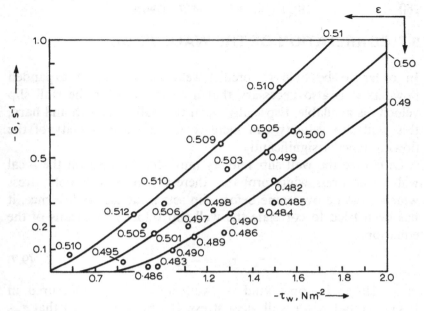

Fig. 9.4. Comparison of experimental (G, τ_w) values with the theoretical relationship for $\epsilon = 0.51$ to $\epsilon = 0.49$.

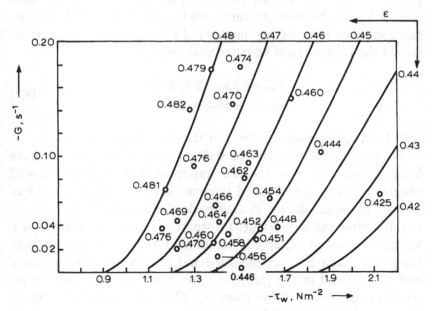

Fig. 9.5. As Fig. 9.4 but for $\epsilon = 0.48$ to $\epsilon = 0.42$.

9.7 CORRELATION OF THE WALL VELOCITY

In order to be able to predict velocity profiles of expanded powders it is also necessary that a correlation for the wall slip velocity is available. Especially when the wall is smooth and hard, this wall slip velocity can increase the average velocity of the flowing powder significantly.

Of course the wall slip velocity must depend also on the local wall shear stress while probably there will be a wall yield stress which must be overcome before slip sets in at the wall. Hence, it has been tried to correlate the wall slip velocity by means of the equation

$$\tau_w - \tau_{w0} = -\mu_w v_w \tag{9.7}$$

in which μ_w is some kind of 'wall-slip-viscosity' (measured in $N s m^{-3}$) and τ_{w0} a wall yield stress. It can be expected that τ_{w0} and perhaps also μ_w will depend on the wall roughness and on the porosity of the powder flowing along the wall. In Fig. 9.6 v_w is plotted versus τ_w. It can be seen that there is a reasonable correlation in the porosity range $0.45 < \varepsilon < 0.50$ but that the correlation is poor at higher porosity. In the porosity range mentioned the results could be correlated well by

$$\left. \begin{array}{l} -\tau_{w0} = 6.84 - 12.7\varepsilon \; N \; m^{-2} \\ \mu_w = 78.5\varepsilon - 28.9 \; N \; s \; m^{-3} \end{array} \right\} \tag{9.8}$$

with both correlations holding for the smooth glass pipe.

To investigate the effect of wall roughness, experiments have been carried out in a perspex standpipe with its internal wall roughened by a thin layer of silicone rubber to which a layer of powder particles adheres. A drawback of this measure, however, is that the pipe wall is no longer transparent and, hence, the wall slip velocity cannot be observed. Assuming, however, that in the bulk the powder flow behaves the same and that the correlations for μ and τ_0 as found in the glass pipe also hold for the perspex pipe, while τ_w now also follows from eqn (9.1), the quantity G can be determined from eqn (9.4). Again $\bar{v}_d$ is determined experimentally so that v_w can be calculated from eqn (9.3) and next τ_{w0} from eqn

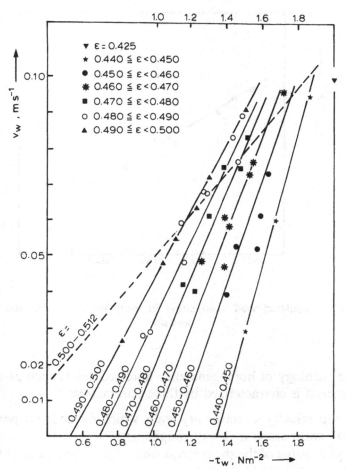

Fig. 9.6. Correlation of wall slip velocity v_w with wall shear stress τ_w with ϵ as parameter. Results from glass tube measurements.

(9.7). It was found that the wall yield stress is about seven times as high as in the glass pipe, see Fig. 9.7.

9.8 CONCLUSIONS

Stable, non-bubbling flow of fluidized solids in a vertical standpipe at various porosities can be achieved by careful control of gas transport and pressure gradient.

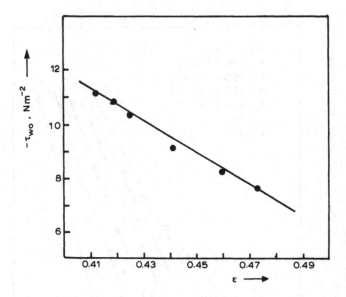

Fig. 9.7. Calculated wall yield stress in roughened perspex pipe versus porosity.

The rheology of homogeneously fluidized solids such as cracking catalyst is characterized by three parameters:

(1) a slip velocity at stationary walls which depends on porosity, wall shear stress and wall roughness;
(2) a yield value of the shear stress which has to be exceeded before shear sets in; and
(3) a powder viscosity.

Yield value and powder viscosity both depend on the porosity. Further studies with other powders are necessary to complete the rheological picture.

REFERENCES

Ashwin, B.S., Hagyard, T., Saunders, I.C.B & Young, T.E. (1960). Viscometers having damped torsional oscillation. *J. Sci. Instr.*, **37**, 480.

Bessant, D.J. (1973). The flow properties of fluidized solids. PhD Thesis, Birmingham, p. 31.

Bessant, D.J. & Botterill, J.S.M. (1973). The flow properties of fluidized solids. *Proc. Int. Symp. on Fluidization and its Applications*, Toulouse, p. 81.

Botterill, J.S.M. & Abdul Halim, B.H. (1978). The flow of fluidized solids. *Proc. 2nd Eng. Found. Conf. on Fluidization*, Cambridge, p. 122.

Botterill, J.S.M. & Abdul Halim, B.H. (1979). The open-channel flow of fluidized solids. *Powder Technol.*, **23**, 67.

Botterill, J.S.M. & Bessant, D.J. (1973). The flow properties of fluidized solids. *Powder Technol.*, **8**, 213.

Botterill, J.S.M. & Bessant, D.J. (1975). The flow properties of fluidized solids. *Fluidization Technology*, vol. II, ed. D.L. Kearns, Hemisphere Publ. Corp., Washington, p. 7.

Botterill, J.S.M. & Van der Kolk, M. (1971). The flow properties of fluidized solids. *AIChE Symp. Ser.*, **67**, 70.

Botterill, J.S.M., Van der Kolk, M., Elliott, D.E. & McGuigan, S. (1972). The flow of fluidized solids. *Power Technol.*, **6**, 343.

Diekman, R. & Forsythe, W.L. (1953). Lab. prediction of flow properties of fluidized systems. *Ind. Eng. Chem.*, **45**, 1174.

Fa-Ke Liu, F. & Orr, C. (1960). Apparent viscosity of gas–solid fluidized systems. *J. Chem. Eng. Data.*, **5**, 430.

Furukawa, J. & Omae, T. (1958). Liquid-like properties of fluidized systems. *Ind. Eng. Chem.*, **50**, 821.

Grace, J.R. (1970). The viscosity of fluidized beds. *Can. J. Chem. Eng.*, **48**, 30.

Hagyard, T. & Sacerdote, A.M. (1966). Viscosity of suspensions of gas-fluidized spheres. *I. and E.C. Fundamentals*, **5**, 500.

Ishida, M., Hatano, H. & Shirai, T. (1980). The flow of solid particles in aerated inclined channels. *Powder Technol.*, **27**, 7.

Kramers, H. (1951). On the 'viscosity' of a bed of fluidized solids. *Chem. Eng. Sci.*, **1**, 35.

Lehnann, J., Ritzmann, H. & Schügerl, K. (1973). Influence of scaling up and type of gas distributor on the behaviour of fluidized beds. *Proc. Int. Symp. on Fluidization and its Applications*, Toulouse, p. 107.

Leont'ev, A.P. & Vakhrushev, I.A. (1976). Experimental determination of the effective viscosity of fluidized beds by the falling ball method. *Khim. Technol. Topl. Masel.*, **4**, 35.

Matheson, G.L., Herbst, W.A. & Holt, P.H. (1949). Fluid solid systems *Ind. Eng. Chem.*, **41**, 1099.

McGuigan, S.J. & Pugh, R.P. (1976). The flow of fluidized solids in an open channel. *Pneumotransport 3*, Bath, E2.

Mori, Y., Aoki, R., Oya, K. & Ishiwaka, H. (1955). Transportation of solid materials by an air-slide conveyor. *Kagaku Kogaku*, **19**, 16.

Muskett, W.J., Leicester, A.R. & Mason, J.S. (1973). The fluidized transport of powdered materials in an air-gravity conveyor. *Pneumo-transport 2*, Guilford F1.

Mutsers, S.M.P. & Rietema, K. (1977). The effect of interparticle forces on the expansion of a homogeneously gas-fluidized bed. *Powder Techn.*, **18**, 239.

Neuzil, L. & Turcajova, M. (1969). Fluidized bed viscosity. *Coll. Czechoslov. Chem. Commun.*, **34**, 3652.

Neuzil, L. & Turcajova, M. (1977). Relative viscosity of fluidized beds. *Coll. Czechoslov. Chem. Commun.*, **42**, 599.

Omae, T. & Furukawa, J. (1954). Rheological properties of fluidized particles. *Kogyo-Kagaku Zasshi*, **57**, 788.

Prudhoe, J. & Whitmore, R.L. (1964) Terminal velocity of spheres in fluidized beds. *Brit. Chem. Eng.*, **9**, 371.

Rietema, K. & Mutsers, S.M.P. (1973). The effect of interparticle forces on the expansion of a homogeneously gas-fluidized bed. *Proc. Int. Symposium on Fluidization and its Applications*, Toulouse. Cepadues-Editions, Carnot France, p. 28.

Ritzmann, H. & Schügerl, K. (1974). The influence of the scale up on the local rheological properties of gas fluidized beds. *Chem. Eng. Sci.*, **29**, 427.

Schenk, G. & Rietema, K. (1979). Viscosity in fluidized beds. Internal report, Eindhoven University of Technology.

Schügerl, K. (1973). Rheological behaviour of fluidized systems. In *Fluidization*, ed. Davidson and Harrison, Academic Press, London, p. 261.

Schügerl, K., Merz, M. & Fetting, F. (1961). Rheologische Eigenschaften von gasdurchströmten Fliessbettsystemen. *Chem. Eng. Sci.*, **15**, 1.

Shinohara, K., Sarto, K., Tanaka, T. (1974). *Micromeretics*, **19**, 64.

Siemens, W. & Hellmer, L. (1962) Die Messung der Wirbelschicht-viscosität mit der pneumatischen Rinne. *Chem. Eng. Sci.*, **17**, 555.

Singh, B., Callcott, T.G. & Rigby, G.R. (1978). Flow of fluidized solids and other fluids in open channels. *Powder Technol.*, **20**, 99.

Trawinski, H. (1953). Effektive Zähigkeit und Inhomogenität von Wirbel-schichten. *Chemie-Ing. Techn.*, **25**, 229.

Turcajova, M. & Neuzil, L. (1976). Fluidized bed flow in an airslide. *Sci. papers Prague Inst. of Chem. Techn.*, 39.

Turcajova, M. & Neuzil, L. (1977). Friction factor for flow of a fluidized bed in an airslide. *Coll. Czechoslov. Chem. Commun.*, **42**, 612.

Van den Langenberg-Schenk, G. & Rietema, K. (1984). The rheology of gas fluidized solids, studied in a vertical standpipe. *Powder Techn.*, **38**, 23.

Van den Langenberg-Schenk, G. (1982). The rheology of gas fluidized powders as determined in a vertical standpipe. PhD thesis, Technical

University of Eindhoven, Eindhoven, The Netherlands.

Van der Leeden, P. & Bouwhuis, G.J. (1960). Tentative rules for shearing stresses in particulate fluidized beds. *Appl. Sci. Res.*, **10**, 78.

Woodcock, C.R. & Mason, J.S. (1976). Fluidized bed conveying. Art or Science. *Pneumotransport 3*, Bath, E1.

Woodruff, H.C. (1973). Rheological properties of plastic–mineral composite powders affecting fluidization and air transport. *Powder Technol.*, **8**, 283.

9.A APPENDIX

The volumetric powder flow rate in the pipe is given by

$$Q_d = 2\pi \int_0^R (1-\varepsilon) r v(r) \, dr$$

Assuming $\partial\varepsilon/\partial r = 0$ and using the boundary condition $v = v_w$ at $r = R$ one finds that

$$Q_d = \pi(1-\varepsilon)\left(R^2 v_w - \int_0^R r^2 \frac{dv}{dr} \, dr \right)$$

The relation for the shear stress in a vertical circular tube reads

$$\tau = -\frac{r}{2}\left\{ \frac{dp}{dz} - \rho_d(1-\varepsilon)g \right\}$$

which implies that

$$dr = \frac{-2d\tau}{(dp/dz) - \rho_d(1-\rho)g}$$

(dv_r/dr) is a function of the shear stress: $(dv_r/dr) = f(\tau)$. Hence, Q_d can be expressed by

$$Q_d = \pi(1-\varepsilon)\left\{ R^2 v_w - \left(\frac{R}{\tau_w}\right)^3 \int_0^{\tau_w} \tau^2 f(\tau) \, d\tau \right\} \qquad (9.A.1)$$

Defining

$$G = \frac{\bar{v}_d - v_w}{R} = \frac{Q_d}{\pi R^3} - \frac{v_w}{R}$$

one finds that

$$G = -\frac{1}{\tau_w^3} \int_0^{\tau_w} \tau^2 f(\tau)\, d\tau \qquad (9.A.2)$$

Differentiation gives

$$\frac{dG}{d\tau_w} = \frac{3}{\tau_w^4} \int_0^{\tau_w} \tau^2 f(\tau)\, d\tau - \frac{1}{\tau_w}\left(\frac{dv}{dr}\right)_w = -\frac{3}{\tau_w}G - \frac{1}{\tau_w}\left(\frac{dv}{dr}\right)_w \qquad (9.A.3)$$

from which

$$-\left(\frac{dv}{dr}\right)_w = 3G + \tau_w \frac{dG}{d\tau_w} \qquad (9.A.4)$$

According to the Bingham model with slip at the wall,

$$f(\tau) = \frac{dv}{dr} = (\tau_0 - \tau)/\mu \qquad \text{for } r > r_0$$

$$v = \text{const.} \qquad\qquad \text{for } r < r_0$$

$$v = v_w \qquad\qquad\quad\ \text{for } r = R$$

With $\tau = (r/R)\tau_w$ and $\tau_0 = (r_0/R)\tau_w$ it now follows from eqn (9.A.2) that

$$G = -\frac{\tau_w}{\mu R^4} \int_0^R r(r_0 - r)\, dr \qquad (9.A.5)$$

On elaborating it is found that

$$G = \frac{\tau_w}{4\mu} + \frac{\tau_0^4}{12\mu\tau_w^3} - \frac{\tau_0}{3\mu} \qquad (9.A.6)$$

10

Fluidization in Practice

NOTATION

a	$= 1/\tau_r$ (s^{-1})
A	Cross-sectional area of fluidized bed (m^2)
b	See Fig. 10.3 (m)
B	See eqn (10.28) (s)
d_p	Particle diameter (m)
D	Bed diameter (m)
D_b	Diameter of gas bubble (m)
F	Slip force (N m^{-3})
g	Gravitational acceleration constant (m s^{-2})
G	Total gas throughput $(\text{m}^3 \text{ s}^{-1})$
h_b	See Fig. 10.3 (m)
H	Bed height (m)
H_b	Bed height before collapse (m)
H_e	$(1-\delta)H_b$ (m)
H_0	Bed height of packed bed (m)
I_d	See eqn (10.26) $(\text{m}^2 \text{ s}^{-1})$
K	Constant introduced by Rowe and Partridge (1965) (—)
l	Coordinate along a streamline (m)
m	Gas mobility in dense phase $(\text{N m}^4 \text{ s}^{-1})$

M	Total bubble volume contents of fluidized bed (m^3)
n_b	Number of bubbles detected per unit time (s^{-1})
N_b	Bubble frequency (s^{-1})
p	Pressure in gas phase (N m^{-2})
p_0	Pressure inside gas bubble (N m^{-2})
p_1	Pressure at height z_1 far above a bubble (N m^{-2})
Q	Total gas flow through gas bubble (m^3 s^{-1})
r	Radial coordinate (m)
R	Radius of gas bubble (m)
R_a	Radius of fluidized bed (m)
R_c	Radius of cloud (m)
R_0	Radius of annulus (m)
S	See eqn (10.17) (m^2 s^{-2})
t	Time (s)
U	Propagation velocity of a disturbance (m s^{-1})
U_b	Rising velocity of gas bubble (m s^{-1})
U_c	Velocity of a continuity wave (m s^{-1})
U_d	Velocity of a dynamic wave (m s^{-1})
v_{b0}	Superficial velocity of gas bubbles (m s^{-1})
v_c	Linear gas velocity in dense phase (m s^{-1})
v_{co}	Superficial gas velocity through bed (m s^{-1})
v_d	Linear velocity of powder particle (m s^{-1})
v_{da}	Velocity of dense phase in the annulus (m s^{-1})
v_{dc}	Velocity of dense phase in the core (m s^{-1})
v_{mf}	Minimum fluidization velocity (m s^{-1})
v_s	Slip velocity $= v_d - v_c$ (m s^{-1})
V_b	Volume of a gas bubble (m^3)
z	Vertical coordinate (m)
z_0	Height inside dense phase where pressure $= p_1$ (m)
α	$= U_b/\varepsilon v_c$ (—)
γ	$= h_b/R$ (—)
δ	Fractional bubble hold-up (—)
δ^*	Local bubble hold-up (—)
δ_0	$= V_{co}/U_b$ (—)
ε	Porosity of a powder (—)
ε_d	Porosity of dense phase (—)

ε_{mf}	Porosity at minimum fluidization (—)
ε_0	Porosity of homogeneous powder bed (—)
θ	Angle in cylindrical coordinates (—)
θ_1	Wake angle of a rising bubble, see Fig. 10.3 (—)
λ	Circulation parameter $= 4\mu_d v_{c0}/[(1-\varepsilon)\rho_d g R_a^2]$ (—)
μ	Gas viscosity (N s m^{-2})
μ_d	'Viscosity' of dense phase (N s m^{-2})
ρ_c	Density of gas (kg m^{-3})
ρ_d	Density of powder particles (kg m^{-3})
τ_d	Shear stress tensor of dispersed phase ($\tau_{rr}, \tau_{rz}, \tau_{zz}$) (N m^{-2})
τ_r	Relaxation time of a declining disturbance (s)
τ_w	Shear stress in dense phase at the wall (N m^{-2})
ϕ	$(R_0/R_a)^2$ (—)
ψ_c	Stream function of continuous phase (m^2 s^{-1})
ψ_d	Stream function of dispersed phase (m^2 s^{-1})
ψ_s	Stream function of slip velocity (m^2 s^{-1})

10.1 INTRODUCTION

In the fundamental part of this book we mainly dealt with homogeneous fluidization and with the critical conditions for its achievement. The most important restriction we found was that there is an upper limit to the superficial fluidization velocity v_{c0}. For fine powders this upper limit is so low (0·02m s^{-1}) that in the practice of gas-fluidization it is far exceeded so that homogeneous fluidization is seldom achieved. Even when in small-scale laboratory experiments homogeneous fluidization is normal, this will no longer be so after scaling up to practical dimensions, because for most processes (chemical or physical) the residence time of the gas in the powder bed must be limited.

Of course, one could confine the scaling up to enlarging the horizontal dimensions while keeping the height of the bed constant, but the resulting fluidization apparatus would take up excessively large floor space. This would be very unattractive from both an economic and a mechanical point of view. Generally, it is preferred to keep the ratio of bed height to bed diameter (H/D) not smaller than unity. This means that the superficial gas velocity must increase in the same ratio as the bed height. Hence the transition of homogeneous into heterogeneous fluidization is inevitable.

When the critical superficial gas velocity for homogeneous fluidization is exceeded, the powder bed will segregate into a dense phase of powder with a porosity somewhat lower than the maximum porosity obtainable in homogeneous fluidization, and a bubble phase consisting of gas voids—generally called bubbles—that contain hardly any powder and rise fast. In that case most of the gas supplied to the bed will pass through it via these bubbles.

At first sight one tends to conclude that this bubble phase constitutes a strong short-circuiting of the powder bed. Fortunately this is not correct as there is considerable gas exchange between the bubble phase and the dense phase by diffusion and convection.

10.2 THE FREELY BUBBLING BED

In Fig. 10.1 the bed height and the pressure drop over the bed are plotted versus the superficial gas velocity v_{co}. At gas velocities

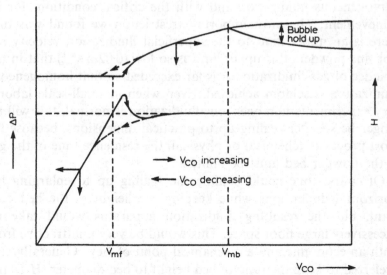

Fig. 10.1. Characteristics of a fluidized bed.

below the incipient gas velocity the bed height remains constant at the packed bed height while the pressure drop continuously increases. If the bed consists of a fine powder this increase is linear. We define a Reynolds number as

$$Re_p = \rho_c d_p v_c / \mu$$

in which ρ_c is the density of the gas, μ is the viscosity of the gas, d_p is the average particle size and v_c the linear gas velocity in the dense phase. If this Reynolds number is larger than 0.1 the increase of the pressure drop will be larger than linear. In this chapter (as in the whole book) we shall deal only with powders for which the Reynolds number will be <0.1 at all gas velocities. In practice this means that the particle size of the powder is $<200\,\mu m$. The pressure gradient over a homogeneous powder bed with a porosity ε_0 is then determined by

$$-\frac{dp}{dz} = \frac{180\mu}{d_p^2} v_{c0} \frac{(1-\varepsilon_0)^2}{\varepsilon_0^3} \quad \text{(Carman, 1937)}$$

When the incipient fluidization velocity v_{mf} is reached the pressure gradient has become $= -(1-\varepsilon_0)(\rho_d - \rho_c)g$. Hence v_{mf} is given by

$$v_{mf} = \frac{(\rho_d - \rho_c)}{180\mu} g d_p^2 \frac{\varepsilon_0^3}{(1-\varepsilon_0)} \quad (10.1)$$

When the gas velocity is further increased the pressure drop will first increase a little more (the surpressure) due to interparticle forces and friction at the wall of the fluidized bed apparatus. At still higher gas velocities this surpressure is soon overcome and the bed expands while the pressure drop remains practically constant at $(1-\varepsilon_0)(\rho_d - \rho_c)gH_0$ where H_0 is the packed bed height. Now the mode of homogeneous fluidization is reached, which is characterized by a uniform porosity all over the bed that increases with increasing gas velocity. At a critical velocity v_{mb} the bed becomes unstable and the first bubbles appear, incidentally at short intervals while the bed height fluctuates. When v_{mb} is exceeded somewhat more, however, the bubbles appear more regularly and uniformly over the cross-section of the bed and the mode of the freely bubbling bed is reached.

When the bed is freely bubbling the total gas flow through the bed is made up of three contributions. Per unit cross-section of the bed, these are:

(1) the gas flow through the dense phase, which equals $(1-\delta)\varepsilon_d v_c$ in which δ is the fractional bubble hold-up, i.e. the fraction of the total bed volume taken in by the bubbles, and ε_d is the porosity of the dense phase;
(2) the total bubble-gas flow, which equals δU_b where U_b is the average rising velocity of the bubbles;
(3) a contribution caused by the fact that the bubbles act as a short-circuit for the gas flow. This contribution equals $3\delta\varepsilon_d v_c$ (see Section 10.3.2).

Hence the total gas flow per unit cross-section is

$$v_{co} = (1 + 2\delta)\varepsilon_d v_c + \delta U_1 \qquad (10.2)$$

10.2.1 The Collapse Experiment

When in a freely bubbling bed the gas supply is suddenly stopped no more bubbles are produced, the last bubbles will leave the bed rapidly and the bed collapses, at first rapidly until all bubbles have left the bed, but then more slowly until also all of the excess gas contained in the dense phase has left the bed and the packed bed height is reached. During this process a series of continuity waves is moving up (see Chapter 6) while at the same time the bed height is decreasing. When the last continuity wave has reached the top of the bed the collapse process is finished (see Fig. 10.2).

This so-called collapse experiment gives us a lot of information on the freely bubbling bed:

(a) the bubble hold-up δ;
(b) the porosity ε_d of the dense phase during free bubbling;
(c) the superficial gas velocity $\varepsilon_d v_c$ through the dense phase during free bubbling.

With regard to (a), when we extrapolate the second part of the curve in Fig. 10.2 towards the moment $(t=0)$ when the gas supply was shut off we find the imaginary bed height H_e that corresponds

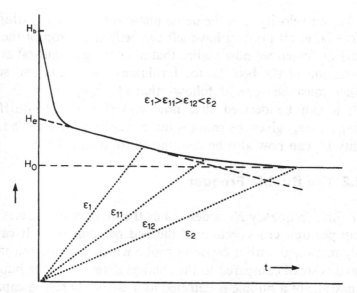

Fig. 10.2. Densification of a fluidized bed (schematically) after shutting off the gas supply.

to the bed contents without the bubbles. Hence the total volume of the bubbles in the bed before collapse is $M = A(H_b - H_e)$ where H_b is the bed height before the collapse. The fractional bubble hold-up δ hence is given by $\delta = (H_b - H_e)/H_b$.

With regard to (b), when ε_0 is the porosity and H_0 the height of the packed bed, the porosity ε_d of the dense phase follows from

$$H_e(1 - \varepsilon_d) = H_0(1 - \varepsilon_0)$$

It appears that ε_d is always somewhat lower than the maximum porosity ε_{ml} of the stable homogeneous fluidized bed. On the other hand ε_d is definitely higher than the packed bed porosity.

As far as (c) is concerned, the linear gas velocity v_c through the dense phase can be calculated from

$$v_c = (1/\varepsilon_d)(-dH/dt)_{\text{no bubbles}} \tag{10.3}$$

The derivation is based on the assumption that the slip velocity $v_s = v_d - v_c$ is a function of the porosity only. Before the collapse, v_d is, on average, zero since the bed is stationary. Hence in that case the slip velocity v_s in the dense phase is $v_{s1} = -v_c$. During the collapse (after all bubbles have left the bed) at the top of the bed $v_d = dH/dt$. When we now realize that after the gas shut-off at any cross-section of the bed the total volume flow of gas and solids together must be zero, it follows that $(1-\varepsilon_d)v_d + \varepsilon_d v_c = 0$, from which it can be derived that now $v_{s2} = (1/\varepsilon_d)v_d = (1/\varepsilon_d)(dH/dt)$. Putting $v_{s1} = v_{s2}$ gives the result wanted. The average bubble rising velocity U_b can now also be determined from eqn (10.2).

10.2.2 The Bubble Frequency

The bubble frequency N_b is defined as the number of bubbles that rise up per unit cross-section of the bed per unit time. It can be simply measured with a capacity probe which is immersed in the bed and is small compared to the average diameter of the bubbles. The passage of a bubble is detected as a slight change in capacity of the probe due to a change of the local dielectric constant.

The number n_b of bubbles detected per unit time is determined by the bubble frequency N_b and the size and shape of the bubbles. To a first approximation we shall assume that all bubbles have the same equivalent diameter D_b (which is the diameter of a perfect sphere with the same volume as the real bubble). The shape of the bubbles is assumed to be that of a spherical cap with radius R, height h_b and base b (see Fig. 10.3). We define $\gamma = h_b/R$ which could vary from 0·5 to 1·6 depending on the powder characteristics and

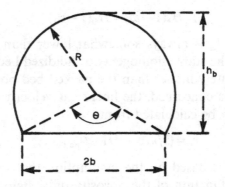

Fig. 10.3. Characteristic dimensions of a rising bubble.

the fluidization rate. Now the volume of the bubbles is given by

$$V_b = \frac{\pi}{6} D_b^3 = \frac{\pi}{3} R^3 \gamma^2 (3 - \gamma)$$

and the relation between R and D_b is found from

$$\left. \begin{array}{c} R = D_b / \sqrt[3]{2\gamma^2(3-\gamma)} \\ \\ b = R\sqrt{\gamma(2-\gamma)} \end{array} \right\} \tag{10.4}$$

while

When $\gamma \leqslant 1$ all bubbles that rise up within a circle with radius b around the probe will be detected, hence in that case

$$\left. \begin{array}{c} n_b = N_b \pi b^2 = N_b \pi R^2 \gamma(2-\gamma) \\ \\ v_{b0} = N_b \dfrac{\pi}{b} D_b^3 = (n_b R) \dfrac{\gamma(3-\gamma)}{3(2-\gamma)} \\ \\ R = (v_{b0}/n_b) \dfrac{3(2-\gamma)}{\gamma(3-\gamma)} \end{array} \right\} \tag{10.5}$$

When on the other hand $\gamma > 1$ then

$$\left. \begin{array}{c} n_b = N_b \pi R^2 \\ \\ v_{b0} = (n_b R) \dfrac{\gamma^2(3-\gamma)}{3} \\ \\ R = (v_{b0}/n_b) \dfrac{3}{\gamma^2(3-\gamma)} \end{array} \right\} \tag{10.6}$$

When v_{b0} is measured by the collapse experiment and n_b with the capacity probe, then N_b and D_b can be found from the above equations:

when $\gamma < 1$ $\quad N_b = (n_b^3/v_{b0}^2) \dfrac{\gamma(3-\gamma)^2}{9\pi(2-\gamma)^3}$

$$D_b = (6v_{b0}/n_b)(2-\gamma)\gamma\{2\gamma^2(3-\gamma)\}^{-2/3} \tag{10.7}$$

when $\gamma > 1$ $N_b = (n_b^3/v_{b0}^2) \dfrac{\gamma^4(3-\gamma)^2}{9\pi}$

$$D_b = (6v_{b0}/n_b)\{2\gamma^2(3-\gamma)\}^{-2/3}$$

(10.8)

In the freely bubbling bed, bubbles grow in size when rising from the bed. Two phenomena are responsible for this growth, as follows.

(1) Since inside a bubble the gas pressure must be the same everywhere while at some distance from the bubble the gas pressure decreases with height, the gas pressure inside the bubble will be some average over the bubble height. Hence gas from the dense phase flows into the bubble at the lower part of the bubble while gas flows out of the bubble in the upper part. Especially when the bubbles are small, more gas flows in than out of the bubble, which means that the bubble must grow in size.

(2) When the bubble hold-up increases, the distance between the bubbles must decrease and the chance that bubbles will coalesce into larger bubbles increases. On the other hand, when the bubbles become too large they become unstable and will split into smaller bubbles by raining of solids through the roof of the bubbles. The larger the bubble the greater the chance that it will split. In particular with fine powders an equilibrium between coalescing and splitting is established, so that higher up in the bed the average bubble size does not grow any more and the bubble frequency, which is highest near the gas distributor, will become constant.

It must be stressed that several variables introduced in the foregoing in fact are average variables: U_b, D_b, n_b, γ. These averages are the result of averaging procedures that are not the same for each of these variables. This means that the equations presented are strictly valid only when the spread of these variables is not too large.

10.3 SINGLE BUBBLES

10.3.1 The Rising Velocity of a Bubble

The rising velocity U_b of a single bubble in a homogeneously gas-fluidized bed has been the subject of research by many investigators, e.g. Harrison and Leung (1962), Davidson and Harrison (1963), Reuter (1965), Rowe and Partridge (1965), Rieke and Pigford (1971), and Rowe and Matsuno (1971).

These authors found that U_b increases with bubble size and varies with the bubble shape, as would be expected. On the basis of a comparison with the potential flow theory of spherical cap bubbles rising in an inviscid liquid, most investigators correlate U_b with the square root of the bubble radius R. Rowe and Partridge (1965) investigated the rising velocity in a three-dimensional bed by means of X-rays. According to them the wake angle θ_b (see Fig. 10.3) of a single bubble varies between 110° and 170° (corresponding to $\gamma = 1.57$ and $\gamma = 1.09$ respectively) and the rising velocity could be correlated by

$$U_b = K\sqrt{gR} \tag{10.9}$$

in which K varied from 0·8 to 1·2. Low values of the wake angle and of the constant K were found with coarse powders and high values with fine powders. In the case of spherical cap bubbles rising in inviscid liquids, experiments show that $\theta_b = 260°$ and $K = 0.67$.

We shall now analyse the solids flow around a bubble in a coordination system that moves up with the bubble. In that case the flow pattern around the bubble can be considered as stationary. The momentum equation of the solids reads

$$(1-\varepsilon)\rho_d \mathbf{v}_d \cdot \nabla \mathbf{v}_d + (1-\varepsilon)\nabla p + \nabla \cdot \tau_d - (1-\varepsilon)\rho_d \mathbf{g} + \mathbf{F} = 0 \tag{10.10}$$

and that of the gas phase

$$\varepsilon \nabla p - \mathbf{E} = 0 \tag{10.11}$$

In these equations τ_d is the tensor representing the forces that the particles exert upon each other by friction and cohesion when they are not free floating; v_d is the velocity vector of the particles, and $\bar{\rho} = (1 - \bar{\varepsilon})\rho_d$.

Elimination of the slip force F and use of the vector equation

$$v_d \cdot \nabla v_d = \tfrac{1}{2} \nabla v_d^2 - [v_d \times \nabla \times v_d] \qquad (10.12)$$

gives

$$\tfrac{1}{2} \nabla v_d^2 - [v_d \times \nabla \times v_d] + \frac{\nabla p}{\bar{\rho}} + \frac{\nabla \cdot \tau_d}{\bar{\rho}} - g = 0 \qquad (10.13)$$

We now integrate this equation scalarly along a streamline l of the solids between two points l_0 and l_1, which correspond to the heights z_0 and z_1 respectively. The second term now disappears since this is a vector which is always perpendicular to the vector v_d and, hence, to the streamline. We find that

$$\frac{1}{2} \int_{l_0}^{l_1} \nabla v_d^2 \cdot dl + \int_{l_0}^{l_1} \frac{\nabla p}{\bar{\rho}} \cdot dl + \int_{l_0}^{l_1} \frac{\nabla \cdot \tau_d}{\bar{\rho}} \cdot dg - \int_{l_0}^{l_1} 1 \cdot dl = 0 \qquad (10.14)$$

We now apply eqn (10.14) along two streamlines: the streamline $\psi = 0$ just above the bubble and the streamline $\psi = \infty$ at a great distance of the bubble (see Fig. 10.4). Both these streamlines run vertically everywhere.

Inside the bubble the pressure is p_0. Along the streamline $\psi = \infty$ this pressure is reached at a height z_0. At a height z_1 far above the bubble the pressure will everywhere be p_1. At the top of the bubble the velocity of the particles is zero. Along the streamline $\psi = \infty$ the velocity is everywhere $-U_b$ which is also the velocity at streamline $\psi = 0$ at height z_1.

Along streamline $\psi = 0$ it is found that

$$\tfrac{1}{2} U_b^2 + \frac{p_1 - p_0}{(1 - \bar{\varepsilon})\rho_d} + \int_{z_0}^{z_1} \frac{[\nabla \cdot \tau_d]_{\psi=0}}{(1 - \varepsilon)\rho_d} \cdot dz + g(z_1 - z_0) = 0 \qquad (10.15)$$

Along streamline $\psi = \infty$ the divergence of the solid shear stress disappears because the velocity and, hence, the stress τ_d are everywhere the same, while the porosity is ε_0. Along this streamline

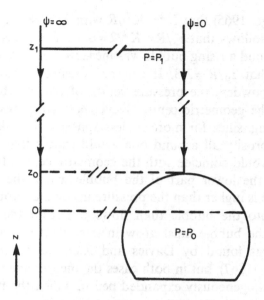

Fig. 10.4. Diagram illustrating the analysis of the rising velocity of a bubble.

we find therefore that

$$\frac{p_1 - p_0}{(1 - \varepsilon_0)\rho_d} + gz_1 = 0 \tag{10.16}$$

Combination of eqns (10.15) and (10.16) gives

$$U_b^2 - 2gz_0 + 2\frac{p_1 - p_0}{\rho_d}\left(\frac{1}{1 - \bar{\varepsilon}} - \frac{1}{1 - \varepsilon_0}\right) + 2S = 0 \tag{10.17}$$

in which

$$S = \int_{z_0}^{z_1} \frac{[\nabla \cdot \tau_d]_{\psi=0}}{(1 - \varepsilon)\rho_d} \cdot \mathrm{d}z$$

Three cases will be discussed:

(1) When the solids are free floating there is no interaction between the individual particles, and the integral S will be zero. When the porosity is everywhere the same the third term in eqn (10.17) will be zero as well and then it is found that $U_b^2 = 2gz_0$. We compare this with the finding of Rowe and

Partridge (1965) that $U_b^2 = K^2 gR$ with $K = 0.8$ for coarse powders. It follows that $z_0/R = K^2/2 = 0.32$. The pressure distribution around a rising bubble was measured by Reuter (1963). He found that $z_0/R = 2/3$. It can be concluded that in beds of coarse powders the pressure centre of rising bubbles is lying above the geometric centre. Nevertheless this result is rather surprising, since for more or less spherical bubbles with constant porosity all around one would expect that the pressure centre would coincide with the geometric centre. It must mean that in the lower part of the bubble where the dense phase pressure is higher than the pressure inside the bubble, more gas flows into the bubble than flows out from the upper part. Hence the bubble must grow in size. That rising bubbles do grow was found by Davies and Richardson (1966) and by Rietema (1967), but in both cases the measurements were done in a homogeneously expanded bed in which the porosity probably was not constant.

(2) When the porosity is not constant throughout the dense phase it will be higher above the bubble, where the distortion of the streamlines will be largest and the third term in eqn (10.17) will be negative. The height z_1 must now be further specified and will be defined as the minimum height above the bubble where the presence of the bubble cannot yet be traced and where the porosity is still equal to ε_0 and hence the third term in eqn (10.17) is still zero. When again the integral S is supposed to be zero (no shear stress in the dense phase, free floating particles), comparison with the result of Rowe and Partridge now shows that z_0 is even smaller than in case (1).

(3) The particles are not free floating, in which case the integral S must be positive. It follows that only this term can compensate for discrepancies in the foregoing analysis.

10.3.2 Gas Exchange with the Dense Phase

As mentioned before, because inside the bubble the pressure is everywhere the same and equal to some average value of the pressure in the dense phase surrounding the bubble, gas must flow into the bubble in the lower part of the bubble and out of the bubble in the upper part. When we assume that the bubble is perfectly spherical and the porosity in the dense phase is everywhere the same, this gas exchange is described by the following analysis.

At constant porosity it follows from the two continuity equations that $\nabla \cdot \mathbf{v}_c = 0$ and $\nabla \cdot \mathbf{v}_d = 0$, hence also $\nabla \cdot \mathbf{v}_s = 0$. The momentum equation of the gas phase (after neglecting the kinetic terms and the shear stress tensor) is

$$\varepsilon \nabla p - \mathbf{F} = 0$$

in which F is the slip force and equals $\mathbf{v}_s/m$ where m is the gas mobility in the dense phase (which is constant when the porosity is constant). From $\nabla \cdot \mathbf{v}_s$ it follows that also $\nabla \cdot \mathbf{F} = 0$ and hence from the momentum equation that

$$\nabla \cdot \nabla p = \nabla^2 p = 0$$

This is the well-known Laplace equation which determines the pressure distribution around the bubble. Together with the boundary conditions it can be solved.

These boundary conditions are:

(1) at the surface of the bubble the pressure is everywhere p_0;
(2) at infinite distance $\partial p/\partial z$ is constant and equal to $-\bar{\rho}g$.

This corresponds with a linear density phase velocity $v_c = \varepsilon_d m \bar{\rho} g$. When the bubble is perfectly spherical the solution of the differential equation runs

$$p - p_0 = \frac{v_{co}}{\varepsilon_d^2 m} \left(r - \frac{R^3}{r^2} \right) \cos \theta \qquad (10.18)$$

The slip velocity can now be found everywhere from the momentum equation. The flow pattern which follows is indicated in Fig. 10.5. By integration of the vertical component of the slip velocity along the surface of the upper half of the bubble, the total gas flow through the bubble can be calculated. This results in $Q_b = 3\pi R \varepsilon_d v_c$ which is three times as much as it would be if the bubble were filled with dense phase of the same porosity. See also eqn (10.2).

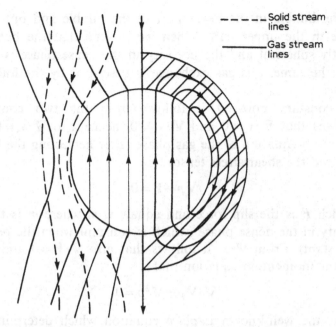

Fig. 10.5. Streamlines of gas and particles through and around a rising bubble.

10.4 THE BUBBLE-CLOUD MECHANISM

The gas that leaves the bubble in the upper part meets there with the oncoming stream of solids (remember that we are still analysing the gas exchange in a coordinate system that moves up with the bubble). The gas tries to penetrate into this solids stream which in its turn tries to entrain the gas leaving the bubble. This, however, is only possible if the solids velocity is higher than the gas velocity. In this connection a parameter α is introduced which equals U_b/v_c. When $\alpha > 1$ the gas will be entrained, which means that the porosity of the solids stream must increase. The entrained gas flows downwards along the bubble sides and reaches the lower region where the gas returns to the bubble. Thus a region exists around the bubble in which the gas is in permanent convective exchange with the bubble gas. This region is called the cloud. The larger α, the thinner is the cloud.

The real existence of this cloud has been beautifully demonstrated by Rowe *et al.* (1964). They worked with a two-dimensional fluidized bed apparatus fitted with glass walls so that two-dimensional rising bubbles could be well observed. Beds of various white powders were fluidized homogeneously with air of 50% humidity and maintained at a temperature of 50°C. Single bubbles of NO_2 gas were injected in the bed. Because of the dark colour of the NO_2 gas the rising bubbles together with their clouds could be readily followed.

Davidson (1961) analysed the phenomenon theoretically on the basis of potential flow of the solids around the bubble. He predicted the cloud radius to be

$$R_c = R \left(\frac{\alpha + 2}{\alpha - 1} \right)^{1/3} \tag{10.19}$$

The experiments of Rowe, however, show that the real cloud thickness is substantially smaller, in particular around the lower part of the bubble. Also the concept of potential flow is questionable because:

(1) it neglects the existence of shear stresses in the solids flow; and
(2) the pressure distribution around the bubble as predicted by potential flow is incompatible with the fact that inside the bubble the gas pressure is everywhere the same.

Nevertheless, it can be shown that, whatever the rheology of the solids, there must be cloud when $\alpha > 1$ (Rietema, 1979).

10.5 GENERAL PROOF OF CLOUD MECHANISM

In connection with this proof we must introduce the concept of stream functions. These are related to the concept of streamlines, which are the paths along which a small element of fluid moves. Hence streamlines never intersect. The connection between stream functions and streamlines is such that to each streamline a value ψ of the stream function is attributed which is constant along the streamline. When we consider two streamlines the difference between their stream functions is a measure of the total amount of fluid that flows in between them. Stream functions and streamlines are only applicable when the flow pattern is stationary and can be described by means of not more than two coordinates. In two-phase systems a stream function can be attributed to each phase.

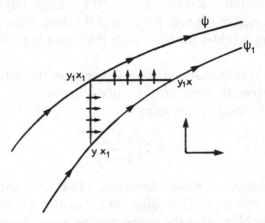

Fig. 10.6. Diagram illustrating the derivation of the stream function.

In $X-Y$ coordinates (see Fig. 10.6) it is found for the continuous phase that

$$\psi_c - \psi_{c1} = -\int_{y_1}^{y} \varepsilon v_{cx} \, dy = \int_{x_1}^{x} \varepsilon v_{cy} \, dx \qquad (10.20)$$

and for the dispersed phase that

$$\psi_d - \psi_{d1} = -\int_{y_1}^{y} (1-\varepsilon)v_{dx} \, dy = \int_{x_1}^{x} (1-\varepsilon)v_{dy} \, dx \qquad (10.21)$$

It follows that

$$\left.\begin{aligned} v_{cx} &= -\frac{1}{\varepsilon}\frac{\partial \psi_c}{\partial y}, \quad v_{cy} = \frac{1}{\varepsilon}\frac{\partial \psi_c}{\partial x} \\ v_{dx} &= -\frac{1}{1-\varepsilon}\frac{\partial \psi_d}{\partial y}, \quad v_{dy} = \frac{1}{1-\varepsilon}\frac{\partial \psi_d}{\partial x} \end{aligned}\right\} \qquad (10.22)$$

Consequently when the densities may be considered constant the continuity equations are automatically satisfied:

$$\frac{\partial}{\partial x}(\varepsilon v_{cx}) + \frac{\partial}{\partial y}(\varepsilon v_{cy}) = 0$$

$$\frac{\partial}{\partial x}\{(1-\varepsilon)v_{dx}\} + \frac{\partial}{\partial y}\{(1-\varepsilon)v_{dy}\} = 0 \tag{10.23}$$

Also in cylindrical coordinates without θ-dependence the stream functions can be defined as:

$$v_{cr} = -\frac{1}{\varepsilon r}\frac{\partial \psi_c}{\partial z}, \quad v_{cz} = \frac{1}{\varepsilon r}\frac{\partial \psi_c}{\partial r}$$

$$v_{dr} = -\frac{1}{(1-\varepsilon)r}\frac{\partial \psi_d}{\partial z}, \quad v_{dz} = \frac{1}{(1-\varepsilon)r}\frac{\partial \psi_d}{\partial r} \tag{10.24}$$

To analyse the bubble-cloud problem we also introduce a stream function of the slip velocity:

$$v_{sr} = -\frac{1}{\varepsilon r}\frac{\partial \psi_s}{\partial z}, \quad v_{sz} = -\frac{1}{\varepsilon r}\frac{\partial \psi_s}{\partial r} \tag{10.25}$$

At any arbitrary height z_1 inside the bubble it follows from $v_s = v_d - v_c$ that

$$\frac{\partial \psi_c}{\partial r} = \frac{\varepsilon}{1-\varepsilon}\frac{\partial \psi_d}{\partial r} - \frac{\partial \psi_s}{\partial r}$$

Integration gives

$$\psi_c = \int_0^r \frac{\varepsilon}{1-\varepsilon}\frac{\partial \psi_d}{\partial r}\,dr - \psi_s = I_d - \psi_s \tag{10.26}$$

in which

$$I_d = \int_0^r \frac{\varepsilon}{1-\varepsilon}\frac{\partial \psi_d}{\partial r}\,dr = \int_0^r \varepsilon r v_{dz}\,dr$$

We now define a function

$$\psi_d^* = -\frac{(\varepsilon_\infty r^2 U_b)}{2}$$

Since $\varepsilon \to \varepsilon_\infty$ and $v_{dz} \to -U_b$ when $r \to \infty$ it follows that $\lim\limits_{r \to \infty} I_d = \psi_d^*$. Hence ψ_d^* is the asymptote of the function I_d.

We also introduce

$$\psi_s^* = -\frac{\varepsilon_\infty r^2 v_{c\infty}}{2}$$

which is the asymptote of ψ_s since $v_{sz} \to v_{c\infty}$ when $r \to \infty$.

We now plot I_d, ψ_d^*, ψ_s and ψ_s^* all at the height z_1 as functions of r^2 (see Fig. 10.7). ψ_d^* and ψ_s^* are both straight lines through the

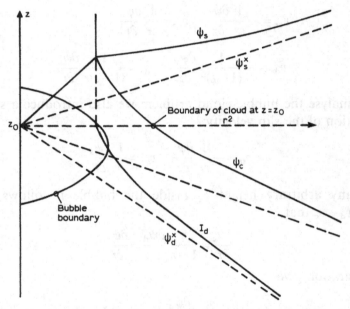

Fig. 10.7. Diagram illustrating the general derivation of the cloud mechanism.

origin. For $r < R$ the function $I_d = 0$ as no solids flow through the bubble void; the slope of I_d is negative. ψ_s is also a curve through the origin and is positive everywhere with a steep slope for $r < R$ since the bubble acts as a short-circuit. Outside the bubble the slope is much smaller as here the slip velocity is limited by the porosity. Besides, ψ_s is everywhere $> \psi_s^*$ and hence $v_{sz} > v_{c\infty}$.

From eqn (10.26) ψ_c can now be derived as a function of r^2. It will be clear that when the slope of $(-\psi_d^*)$ is larger than the slope of ψ_s^* (or when $\alpha = U_b/v_{c\infty} > 1$) there must be a certain radius R_c at which ψ_c is zero.

When the above analysis is repeated at different heights then at each height a specific radius $R_c(z)$ will be found for which $\psi_c = 0$. Since streamlines never intersect, all of these radii will form an envelope around the bubble at which everywhere $\psi_c = 0$. This means that the gas can never pass this envelope (by convection) and hence the existence of a cloud is proved. When $\alpha < 1$ there is no cloud. The larger α, the thinner is the cloud. The real cloud thickness depends also strongly on the solids flow; as this solids flow adheres more strongly to the bubble, so the thinner the cloud will be. In the practice of fluidization of fine powders α will generally be very large and hence the thickness of the cloud can nearly always be neglected.

10.6 CORRELATION WITH INTERPARTICLE FORCES

It has often been argued that in heterogeneous fluidization as it occurs in practical applications, the dense phase is so frequently disturbed by rising bubbles that the theory on bed stability as developed for the case of homogeneous fluidization (Chapters 6, 7 and 8) is of no practical value. It will be shown here that this opinion is based on a lack of understanding of the underlying theory.

As an example we choose the situation in which a cracking catalyst powder is fluidized by argon at a gas pressure of 1 bar and a superficial gas velocity of $0.05\,\mathrm{m\,s^{-1}}$. At this gas rate the dense phase porosity appears to be 0.524 while for this system the bubble point porosity is 0.600 to which belongs a value of the elasticity modulus of $0.278\,\mathrm{N\,m^{-2}}$. The density of the solid amounts to $887\,\mathrm{kg\,m^{-3}}$. From these data it can be calculated that at the bubble point $U_d = U_c = 0.0177\,\mathrm{m\,s^{-1}}$. The bubble frequency was measured in the way indicated in Section 10.2.2 and was found to be $4.3\,\mathrm{s^{-1}}$, which indicates that the time between two bubble passages is $0.23\,\mathrm{s}$.

From the stability theory (Chapter 6) it follows that the rate of decline of a disturbance is given by e^{-at}, where

$$a = \frac{1}{\tau_r} = \frac{B}{2}\left(1 - \frac{U_c}{U}\right) \tag{10.27}$$

where U is the propagation velocity of the disturbance. As U may have any value between U_c and U_d and a is >0

$$B = \frac{180\mu}{\rho_d d_p^2}\left(\frac{1-\varepsilon}{\varepsilon^3}\right) \tag{10.28}$$

From eqns (6.7) and (10.1) it follows that

$$U_c = \frac{\rho_d g d_p^2}{180\mu}\left[\frac{(3-2\varepsilon)\varepsilon^2}{1-\varepsilon}\right] \tag{10.29}$$

The porosity dependence of E is given by

$$E = 0 \cdot 278 \exp[-(\varepsilon - 0 \cdot 60)]$$

Hence, at $\varepsilon = 0 \cdot 524$ these data give

$$U_c = 0 \cdot 0105 \text{ m s}^{-1}$$
$$U_d = 0 \cdot 0313 \text{ m s}^{-1}$$
$$B = 3488 \text{ s}^{-1}$$

It follows that the relaxation time for a disturbance with a propagation velocity only a fraction larger than U_c, e.g. $U = 0 \cdot 0106 \text{ m s}^{-1}$, is already $<0 \cdot 06$ s, while the relaxation time decreases rapidly when the propagation velocity of the disturbance further increases. When, e.g. $U = 0 \cdot 0120 \text{ m s}^{-1}$, τ_r becomes $0 \cdot 005$ s. It can be concluded, therefore, that between two bubble passages the dense phase may be considered to be completely stabilized. Hence, it can be expected that interparticle forces must have influence on other properties of the freely bubbling fluidized bed.

The obvious thing to do, therefore, is to look for a correlation between the elasticity modulus E and, e.g., the average bubble size and the dense phase porosity. In order to have sufficient data available for correlation, we did further experiments with the fluidization equipment that was also used for the experiments on homogeneous fluidization discussed in Chapter 7. Fresh cracking catalyst was fluidized with three different gases, methane, nitrogen and argon, at gas pressures up to 15 bar and at a superficial gas velocity of 0·05 m s^{-1}. The bubble frequency N_b, the dense phase porosity ε_d, the superficial dense phase velocity $\varepsilon_d v_c$ and the bubble hold-up δ were measured. The results are given in Tables 10.1 and 10.2. Table 10.1 presents the measured quantities and Table 10.2 the derived quantities. The bubble radius R was derived from these data with help of eqn (10.7) assuming a value of $\gamma = 1·5$. The parameter K as introduced by Rowe and Partridge (1965) was determined from

$$K = U_b(gR)^{-1/2} \qquad (10.30)$$

The value of E_{50} was taken from the results presented in Chapter 7.

In Fig. 10.8 the bubble radius R is plotted against the elasticity modulus E_{50} with the gas viscosity μ as parameter (for the data on the gas viscosity see Chapter 7). Figure 10.8 strongly suggests a correlation of R with the product of E_{50} and μ as shown in Fig. 10.9. Nevertheless, this strong effect of the gas viscosity on the average bubble radius R is rather surprising.

We suggest that a high viscosity increases the chance that a rising bubble is split at the top of the bubble. This splitting could be caused by a stronger coherence of the stream of solids that approaches the roof of the bubble and tries to penetrate it. This swarm coherence was mentioned in Chapter 1 (Section 1.6). It resists dilution of the swarm since that involves the inflow of gas which is more difficult the higher the gas viscosity. No bubble splitting in fact means that the solids stream approaching from above must be split in order to flow around the bubble. The higher the coherence the more difficult this is and the larger the chance that the bubble will be split, which means a shift of the equilibrium between coalescence and bubble splitting in favour of smaller bubbles.

In Fig. 10.10 the dense phase porosity ε_d is plotted against the product of E_{50} and μ. Here too the results with various gases and various gas pressures combine in a neat curve. This is not surprising since ε_d differs not too much from ε_{mb}—which is also determined by E_{50} and the gas viscosity—albeit in a somewhat different relationship (see Chapter 7).

TABLE 10.1

Measuring Results of Freely Bubbling Bed of Fresh Cracking Catalyst (FCC)

p (bar)	v_{co} (cm s⁻¹)	$\delta(\%)$ CH₄	$\delta(\%)$ N₂	$\delta(\%)$ Ar	v_{bo} (cm s⁻¹) CH₄	v_{bo} (cm s⁻¹) N₂	v_{bo} (cm s⁻¹) Ar	n_b (s⁻¹) CH₄	n_b (s⁻¹) N₂	n_b (s⁻¹) Ar	εd CH₄	εd N₂	εd Ar
1	2	1·4	2·7	3·3	1·43	1·60	1·63	3·05	3·4	3·7	0·358	0·524	0·552
	3	2·8	4·8	5·0	2·51	2·70	2·65	3·16	3·4	3·9	0·501	0·501	0·535
	4	4·7	7·0	6·0	3·54	3·71	3·69	3·06	3·7	4·1	0·489	0·490	0·530
	5	6·9	7·1	7·2	4·53	4·72	4·70	3·06	3·9	4·3	0·480	0·489	0·524
3	2	1·6	3·3	3·1	1·47	1·62	1·59	3·30	3·4	3·7	0·516	0·530	0·560
	3	2·7	4·8	4·9	2·50	2·66	2·60	3·48	4·0	4·2	0·508	0·514	0·549
	4	4·1	6·0	6·2	3·56	3·65	3·64	3·47	4·1	5·0	0·494	0·511	0·545
	5	8·1	7·1	7·4	4·50	4·67	4·61	3·70	4·5	6·1	0·489	0·510	0·544
6	2	2·7	3·7	3·7	1·42	1·57	1·52	3·17	3·3	4·7	0·534	0·541	0·582
	3	3·3	5·7	5·0	2·49	2·60	2·48	3·83	4·4	6·6	0·519	0·534	0·579
	4	4·4	7·3	6·9	3·45	3·59	3·47	3·95	5·0	7·8	0·505	0·534	0·579
	5	8·1	7·8	8·5	4·54	4·60	4·45	4·13	5·7	8·9	0·495	0·531	0·579
9	2	2·4	4·0	3·9	1·34	1·53	1·39	3·46	3·3	6·4	0·540	0·556	0·601
	3	3·8	5·5	5·5	2·44	2·51	2·39	4·00	4·4	8·2	0·523	0·551	0·601
	4	6·3	7·7	7·2	3·46	3·50	3·35	4·21	5·2	10·4	0·514	0·550	0·602
	5	8·3	8·7	8·8	4·50	4·50	4·35	4·69	7·3	11·4	0·504	0·548	0·600
12	2	1·7	3·8	3·7	1·42	1·44	1·30	3·55	4·8	5·9	0·549	0·570	0·618
	3	3·5	5·1	5·5	2·43	2·43	2·27	4·30	7·2	7·5	0·529	0·568	0·618
	4	6·4	7·4	7·5	3·44	3·41	3·26	4·75	8·0	9·7	0·517	0·567	0·618
	5	8·0	9·4	10·4	4·44	4·38	4·22	5·02	10·0	12·9	0·513	0·568	0·618
15	2	1·9	3·5	3·6	1·34	1·38	1·22	3·43	4·6	6·2	0·559	0·582	0·631
	3	5·0	5·6	5·3	2·47	2·35	2·22	4·62	6·8	8·0	5·526	0·580	0·632
	4	6·7	7·2	8·0	3·86	3·42	3·19	5·00	7·5	10·0	0·522	0·579	0·629
	5	7·4	10·3	10·0	4·47	4·31	3·96	5·42	10·2	14·0	0·518	0·578	0·629

TABLE 10.2

Derived Quantities of Freely Bubbling Bed of Fresh Cracking Catalyst (FCC)

p (bar)	v_{co} (cm s^{-1})	$\bar{U}^b$ (cm s^{-1})			R (cm)			K^a			E_{50} (N m^{-2})		
		CH_4	N_2	Ar	CH_4	N_2	Ar	CH_4	N_2	Ar	CH_4	N_2	Ar
1	2	102	59	49	0.42	0.42	0.39	5.03	2.91	2.50	0.99	1.22	1.25
	3	90	56	53	0.70	0.70	0.60	3.43	2.14	2.18			
	4	75	53	62	1.03	0.89	0.80	2.36	1.79	2.21			
	5	66	66	65	1.31	1.07	0.97	1.83	2.04	2.10			
3	2	92	49	51	0.40	0.42	0.38	4.64	2.41	2.64	1.13	1.45	1.74
	3	93	55	53	0.64	0.59	0.55	3.71	2.29	1.96			
	4	87	61	59	0.91	0.79	0.65	2.91	2.19	2.34			
	5	55	66	62	1.08	0.92	0.67	1.69	2.20	2.42			
6	2	53	43	41	0.40	0.42	0.29	2.68	2.12	2.43	1.69	2.25	2.50
	3	75	46	50	0.58	0.52	0.33	3.14	2.04	2.78			
	4	78	49	50	0.77	0.64	0.39	2.84	1.95	2.56			
	5	56	59	52	0.98	0.72	0.44	1.80	2.22	2.50			
9	2	56	38	36	0.34	0.41	0.19	3.07	1.89	2.64	2.23	3.06	3.93
	3	64	46	45	0.54	0.51	0.26	2.78	2.06	2.82			
	4	55	45	47	0.73	0.60	0.29	2.06	1.85	2.79			
	5	54	52	49	0.85	0.55	0.34	2.03	2.24	2.68			
12	2	84	38	35	0.36	0.27	0.20	4.47	2.33	2.50	2.55	3.44	5.34
	3	70	48	41	0.50	0.30	0.27	3.16	2.07	2.52			
	4	54	46	43	0.64	0.38	0.30	2.16	2.38	2.51			
	5	55	47	41	0.79	0.39	0.29	1.98	2.40	2.43			
15	2	70	39	34	0.35	0.27	0.17	3.78	2.40	2.63	2.95	4.50	7.04
	3	49	42	42	0.47	0.31	0.25	2.28	2.41	2.68			
	4	58	47	40	0.68	0.40	0.28	2.24	2.37	2.41			
	5	60	42	40	0.73	0.38	0.25	2.24	2.18	2.55			

^{a}From Rowe & Partridge (1965).

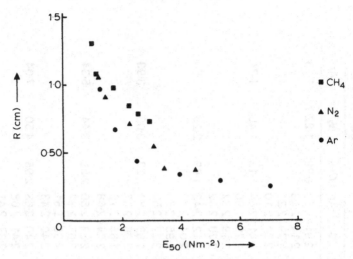

Fig. 10.8. Bubble radius R versus E_{50} for three gases.

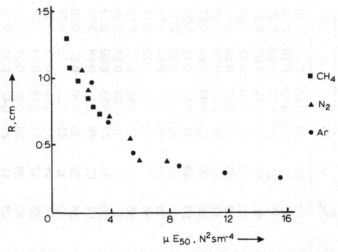

Fig. 10.9. Bubble radius R versus the product of the gas viscosity and E_{50} for three gases.

Finally in Fig. 10.11 the parameter K introduced by Rowe and Partridge is presented as function of the dense phase porosity. It is remarkable that K is about twice as high as the value of K determined by Rowe and Partridge for fine powders. This is

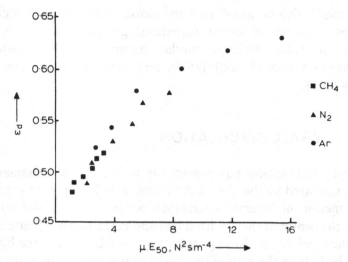

Fig. 10.10. Dense phase porosity ε_d versus μE_{50}.

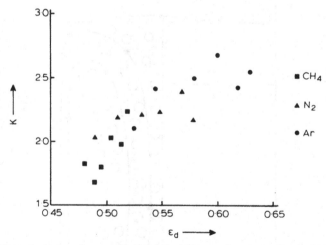

Fig. 10.11. Bubble rising parameter K of Rowe and Partridge (1965) versus the dense phase porosity.

probably due to the fact that Rowe and Partridge carried out their experiments on single rising bubbles in homogeneous fluidized beds at much lower dense phase expansions than were obtained in freely bubbling beds where the term S in eqn (10.17) is much smaller and can nearly be neglected.

It should also be noted that the dense phase porosity tends to become constant at higher superficial gas velocities, as is also shown in Table 10.2. A similar observation was made by Abrahamson and Geldart (1980) and also by Bohle and Van Swaay (1978).

10.7 OVERALL CIRCULATION

In larger fluidization equipment, for which the bed diameter is large compared to the size of the rising bubbles, in most cases the phenomenon of 'overall' circulation occurs: the bubbles tend to rise in the centre of the fluidized bed and entrain dense phase along with them, while the dense phase returns bubble-free to the bottom of the bed along the wall of the fluidized bed apparatus, a so-called bubble street is formed (see Fig. 10.12). This same phenomenon

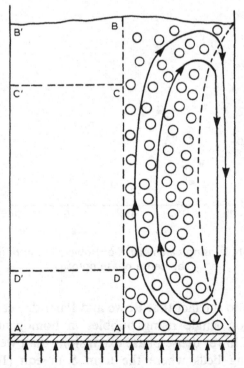

Fig. 10.12. Bubble street formation in a freely bubbling bed.

also occurs in other two-phase systems such as liquid–liquid systems (Wijffels & Rietema, 1972) and liquid–gas systems (Rietema & Ottengraf, 1970). The explanation is based on the fact that overall circulation involves a reduction of the pressure drop over the bed caused by wall friction. The flow pattern becomes such that the energy dissipation will be at a minimum.

10.7.1 The Momentum Equations

The dense phase will be considered as a homogeneous continuous phase of constant density $\bar{\rho}$ with velocity $\mathbf{v}_d$ and shear stress tensor τ. The momentum equations then read (see Chapter 5) for the dense phase:

$$(1-\delta)\bar{\rho}\mathbf{v}_d \cdot \nabla \mathbf{v}_d + (1-\delta)\nabla \cdot \tau + (1-\delta)\nabla p - (1-\delta)\bar{\rho}\mathbf{g} - \mathbf{F} = 0$$

and for the bubble phase:

$$\delta \nabla \cdot \tau + \delta \nabla p + \mathbf{F} = 0$$

Addition gives

$$(1-\delta)\bar{\rho}\mathbf{v}_d \cdot \nabla \mathbf{v}_d + \nabla \cdot \tau + \nabla p - (1-\delta)\bar{\rho}\mathbf{g} = 0 \qquad (10.31)$$

On elaboration in cylindrical coordinates without θ-dependence and with axial symmetry we get for the r-component:

$$(1-\delta)\bar{\rho}\left(v_{dr}\frac{\partial v_{dr}}{\partial r} + v_{dz}\frac{\partial v_{dr}}{\partial z}\right) + \frac{\partial p}{\partial r} + \left(\frac{1}{r}\frac{\partial}{\partial r}r\tau_{rr} + \frac{\partial \tau_{rz}}{\partial z}\right) = 0 \qquad (10.32)$$

and for the z-component:

$$(1-\delta)\bar{\rho}\left(v_{dr}\frac{\partial v_{dz}}{\partial r} + v_{dz}\frac{\partial v_{dz}}{\partial z}\right) + \frac{\partial p}{\partial z} + (1-\delta)\bar{\rho}g$$

$$+ \frac{1}{r}\frac{\partial}{\partial r}(r\tau_{rz}) + \frac{\partial \tau_{zz}}{\partial z} = 0 \qquad (10.33)$$

We shall now first consider the zone DCC'D' where everywhere the radial velocity v_{dr} is zero (see Fig. 10.12). From the continuity equations it follows then that everywhere $\partial v_{dz}/\partial z = 0$. From

$$\tau_{zz} = -2\mu_d \frac{\partial v_{dz}}{\partial z}, \quad \tau_{rz} = -\mu_d\left(\frac{\partial v_{dz}}{\partial r} + \frac{\partial v_{dr}}{\partial z}\right), \quad \tau_{rr} = -2\mu_d \frac{\partial v_{dr}}{\partial r}$$

it then follows that everywhere

$$\tau_{zz} = 0, \ \tau_{rr} = 0, \ \frac{\partial \tau_{rz}}{\partial z} = 0, \ \tau_{rz} = -\mu_{\mathrm{d}} \frac{\partial v_{\mathrm{d}z}}{\partial r}$$

From eqn (10.32) it follows that $\partial p / \partial r = 0$ and from eqn (10.33) that

$$-\frac{\partial p}{\partial z} = (1 - \delta^*) \bar{\rho} g + \frac{1}{r} \frac{\partial}{\partial r} (r \tau_{rz}) \tag{10.34}$$

δ^* is the local bubble hold-up. Inside the bubble street with radius R_0, $\delta^* \neq 0$ but constant.

Integration of eqn (10.34) gives for $0 < r < R_0$

$$-\tau_{rz} = \frac{r}{2} \left\{ \frac{\partial p}{\partial z} + (1 - \delta^*) \bar{\rho} g \right\} \tag{10.35}$$

At $R_0 < r < R_a$ everywhere $\delta^* = 0$. At $r = R_0, \tau_{rz}$ must be continuous, hence after integration it is found for $R_0 < r < R_a$ that

$$-\tau_{rz} = \frac{r}{2} \left(\frac{\partial p}{\partial z} + \bar{\rho} g \right) - \frac{R_0^2}{2r} \delta^* \bar{\rho} g \tag{10.36}$$

The shear stress τ_{w} which the wall exercises on the downward-flowing dense phase is directed upwards and given by

$$\tau_{\mathrm{w}} = (-\tau_{rz})_{r=R_a} = \mu_{\mathrm{d}} \left(\frac{\partial v_{\mathrm{d}z}}{\partial r} \right)_{r=R_a}$$

Note that near the wall $v_{\mathrm{d}z}$ is negative but its absolute value decreases with increasing r. With $\phi = (R_0 / R_a)^2$, eqn (10.36) now gives for $r = R_a$

$$-\frac{\partial p}{\partial z} = (1 - \delta^* \phi) \bar{\rho} g - \frac{2}{R_a} \tau_{\mathrm{w}} \tag{10.37}$$

Hence as a consequence of the circulation the pressure gradient and also the pressure drop over the bed decreases.

Of course eqn (10.37) does not hold in the upper and the lower zones where the velocity of the dense phase reverses. Along B'C' and D'A', however, we can put $v_{\mathrm{d}r} = 0$ and $\partial v_{\mathrm{d}z} / \partial z = 0$ while $v_{\mathrm{d}z} < 0$. Along B'C' $\tau_{zz} < 0$ and $\partial v_{\mathrm{d}z} / \partial z > 0$. Along A'D' $\tau_{zz} > 0$ and $\partial v_{\mathrm{d}z} / \partial z < 0$.

Equation (10.33) reduces to

$$-\frac{\partial p}{\partial z} = \underbrace{(1-\delta^*\phi)\bar{\rho}g}_{\textbf{(a)}} + \underbrace{\frac{1}{r}\frac{\partial}{\partial r}(r\tau_{rz})}_{\textbf{(b)}} + \underbrace{(1-\delta^*)\bar{\rho}v_{dz}\frac{\partial v_{dz}}{\partial z}}_{\textbf{(c)}} + \underbrace{\frac{\partial \tau_{rz}}{\partial z}}_{\textbf{(d)}} \tag{10.38}$$

When now we suppose that the circulation pattern is symmetrical with respect to a horizontal plane at half the height of the bed, then with each slice Δz in the lower zone a slice Δz in the upper zone can be associated where integration with respect to r gives as a result that the contributions to the integral of the terms **(c)** and **(d)** over the upper and the lower zone compensate each other while the terms **(a)** and **(b)** again give eqn (10.37) as a result. This means that when we integrate eqn (10.38) over the full height of the bed we obtain the same result as when we integrate eqn (10.37) over the height. Hence we find that

$$\Delta p = \bar{\rho}gH_e - \bar{\rho}g \int_0^{H_e} \delta^*\phi \, dz - \frac{2}{R_a} \int_0^{H_e} \tau_w \, dz \tag{10.39}$$

In this equation H_e indicates the expanded bed height. The integral in the second term on the right-hand side when multiplied by the cross-sectional area A becomes equal to the total bubble contents M of the bed. Since AH_e is equal to the total bed volume, $(AH_e - M)$ is equal to the total volume of the dense phase and hence $\bar{\rho}g(AH_e - M)$ is the total weight of the bed $= \rho_0gH_0A$. Equation (10.39) can now be written as

$$\Delta p = (\Delta p)_0 - \frac{2}{R_a} \int_0^{H_e} \tau_w \, dz \tag{10.40}$$

in which $(\Delta p)_0$ is the pressure drop over the bed at incipient fluidization.

10.7.2 Derivation of the Circulation Pattern

For the calculation of the circulation we have eqns (10.35) and (10.36) and the relation between the shear stress and the shear rate (the rheology). The unknown variables in these equations amount to seven: the velocity v_{dc} in the bubble street, the velocity v_{da} in the annulus, the pressure gradient $\partial p/\partial z$, the bubble hold-up δ^*, the expanded bed height H_e, the ratio $\phi=(R_0/R_a)^2$ and finally the shear stress τ_w at the wall. We also have three more equations: the dense phase mass balance, the dense phase circulation balance and the gas balance. This means, however, that we are still one equation short and hence there must be a whole series of solutions. The extra information we need to solve the problem is found in the circumstance that every dynamic system—such as the fluidized bed—in which energy is dissipated always strives toward such a configuration that the total energy dissipated will be minimum (see Rietema & Ottengraf, 1970, appendix). Hence when we choose a certain value of, e.g., the bubble hold-up in the bubble street, the other variables can be solved and the energy dissipated ($= G\Delta p$) can be calculated. When this is done for every value of δ^* it will be found that there is a minimum at a certain value of δ^*.

Other authors have also proposed models for the circulation problem but these models are all based on the concept of inviscid flow of the dense phase, which is unrealistic. Besides, these models give no explanation why circulation should occur at all. We now return to our model and will first formulate the three additional balances mentioned before.

Firstly, the mass balance indicates that the total mass in the bed is constant, hence

$$(1-\varepsilon_0)H_0=(1-\varepsilon)\{1-\delta^*\phi\}H_0 \qquad (10.41)$$

The circulation balance of the dense phase says that the total amount of dense phase that moves up in the bubble street must be equal to the amount moving down in the annular zone; hence with $\bar{v}_{dc}$ the average velocity in the core and $\bar{v}_{da}$ the average velocity in the annular zone this gives

$$\phi(1-\delta^*)\bar{v}_{dc}+(1-\phi)\bar{v}_{da}=0 \qquad (10.42)$$

Lastly, the gas balance is somewhat more complicated. If the dense phase were not circulating, the linear velocity of the gas in the bubble phase would equal $U_b+3\varepsilon v_s$. When, however, there is circulation this velocity will be increased with the linear velocity of the dense phase. The total gas throughput via the bubble phase is therefore

$$\delta^*(U_b + 3\varepsilon v_s + \bar{v}_{dc})\pi R_0^2$$

Gas is also transported through the upward-flowing dense phase in the bubble street to the amount of

$$(1 - \delta^*)\varepsilon(v_s + \bar{v}_{dc})\pi R_0^2$$

Finally the downward-moving dense phase in the annular zone transports gas, viz.

$$\varepsilon(v_s + \bar{v}_{da})\pi(R_a^2 - R_0^2)$$

Note that $\bar{v}_{da}$ is negative except near the boundary with the bubble street (see Fig. 10.13). Altogether it is found that

$$\pi R_a^2 v_{co} = \delta^*(U_b + 3\varepsilon v_s + \bar{v}_{dc})\pi R_0^2 \tag{10.43}$$
$$+ (1 - \delta^*)\varepsilon(v_s + \bar{v}_{dc})\pi R_0^2 + \varepsilon(v_s + \bar{v}_{da})\pi(R_a^2 - R_0^2)$$

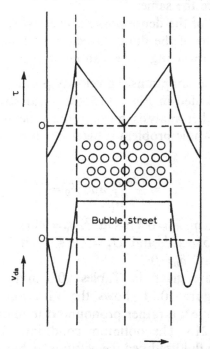

Fig. 10.13. Velocity profile of particles and shear stress profile in a circulating fluidized bed.

After elaboration and combination with eqn (10.42) this result can be reduced to

$$v_{c0} = \delta^* \phi (U_b + \bar{v}_{dc}) + \varepsilon (2\delta^* \phi + 1) v_s \qquad (10.44)$$

For a first, rather simple solution of the circulation problem we made the following additional assumptions:

(1) as circulation mainly occurs at rather high superficial gas velocities and the slip velocity v_s is small in the case of fine powders, we put $v_s = 0$;
(2) the dense phase behaves as a Newtonian liquid, i.e. the viscosity is constant, no slip at the wall and no yield value for the shear stress;
(3) in the annular zone the flow of the dense phase is laminar;
(4) the rising bubbles in the core induce such a strong radial mixing that in this core the vertical velocity of the dense phase is everywhere the same;
(5) the porosity of the dense phase is everywhere the same;
(6) the H/D ratio of the fluidized bed is so high that in this first analysis the reversing zones can be neglected.

With the above assumptions the velocity profile of the dense phase must be as indicated in Fig. 10.13; the shear stress profile is also indicated. A further analysis, given in the appendix to this chapter, shows that the whole problem is determined by two dimensionless parameters, viz.

$$\lambda = \frac{4\mu_d v_{co}}{\bar{\rho} g R_a^2} \text{ and } \delta_0 = \frac{v_{co}}{U_b}$$

With $R_a = 0.6325$ m, $\mu_d = 0.3$ N s m^{-2}, $\rho_d = 900$ kg m^{-3}, $g = 10$ m s^{-2}, $\varepsilon = 0.4$, $v_{co} = 0.5$ m s^{-1} and $U_b = 2$ m s^{-1}, it is found that $\lambda = 0.2778 \times 10^{-3}$ and $\delta_0 = 0.25$.

The results are given in Tables 10.3 and 10.4 and in Figs 10.14–10.18. Figure 10.14 shows the variation of $\Delta p^* = \Delta p/\Delta p_0$ with variation of δ^*; a rather pronounced minimum is shown at a value of $\delta^* = 0.215$. The optimum conditions for Δp at various diameters of the fluidized bed (i.e. various λ), but always at a value of $\delta_0 = 0.25$, are given in Table 10.4 and Figs 10.15–10.18.

TABLE 10.3
Determination of Optimum Conditions (Minimum Δp^*) of Circulating Fluidized Bed While $d\lambda = 0.2778 \times 10^{-3}$ and δ_0 is constant $(=0.25)$

δ^*	φ	β	T_{c4}	T_{c8}	Δp^*
0·15	0·445	0·612 53	0·033 23	0·033 20	0·973 1
0·18	0·582	0·733 05	0·011 75	0·011 67	0·969 6
0·20	0·647	0·783 36	0·006 617	0·006 609	0·968 7
0·212	0·681	0·808 17	0·004 738	0·004 745	0·968 49
0·215	0·689	0·813 86	0·004 362	0·004 366	0·968 49
0·22	0·702	0·823 01	0·003 799	0·003 797	0·968 52
0·23	0·728	0·840 89	0·002 836	0·002 833	0·968 8
0·25	0·781	0·875 78	0·001 429	0·001 423 3	0·970 6

TABLE 10.4
Calculated Circulation Parameters (Minimum Δp^*) with Variation of Bed Diameter (varying λ) While δ_0 is constant $(=0.25)$

$\lambda \times 10^3$	R_a (m)	φ	δ^*	$(1-\Delta p^*) \times 10^3$	$\bar{v}_{dc}$ $(m\,s^{-1})$	$\bar{v}_{da}$ $(m\,s^{-1})$
0·444 4	0·500	0·650	0·230	36·56	1·345	1·923
0·277 8	0·632	0·689	0·214	31·51	1·375	2·391
0·177 8	0·791	0·720	0·203	27·34	1·382	2·914
0·133 3	0·913	0·746	0·198	24·92	1·411	3·324
0·066 66	1·291	0·790	0·185	19·91	1·421	4·357
0·044 44	1·581	0·811 5	0·178	17·44	1·462	5·172

10.7.3 Discussion

The most doubtful assumption made in the foregoing analysis certainly concerns neglecting the reversing zones. The error thus made, however, can be corrected for the greater part by making a rough estimate of the local wall friction which pertains there and accounting for it by means of eqn (10.40). To that end we estimate the height of the reversing zones to be equal to the radius R_a of the fluidized bed. At the top and the bottom of the bed $v_{dz} = 0$ and hence also $\tau_w = 0$. The average value of τ_w therefore can be estimated to be two-thirds of the value which pertains to the

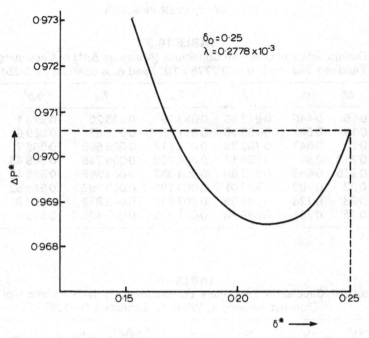

Fig. 10.14. Dimensionless pressure drop Δp^* over the bed as a function of bubble hold-up in the bubble street.

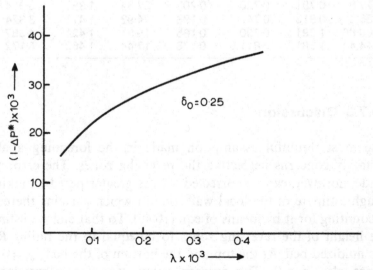

Fig. 10.15. Relative gain of pressure drop over the bed as a function of the scaling parameter λ at $\delta_0 = 0.25$.

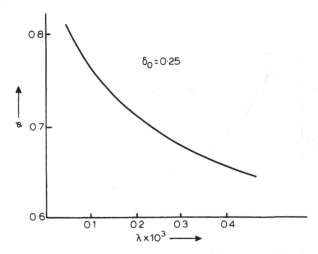

Fig. 10.16. Relative bubble street cross-sectional area versus the scaling parameter $\lambda(\delta_0 = 0.25)$.

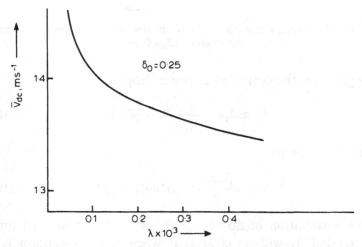

Fig. 10.17. Average particle velocity in the core versus the scaling parameter $\lambda(\delta_0 = 0.25)$.

middle zone. The integral in eqn (10.40) can now be evaluated as

$$\int_0^{H_e} \tau_w \, dz \approx \left(H_e - \frac{2}{3} R_a \right) \tau_{wm}$$

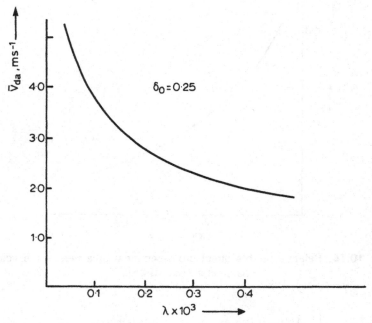

Fig. 10.18. Average particle velocity in the annulus versus the scaling parameter $\lambda (\delta_0 = 0.25)$.

This gives for the corrected pressure drop

$$\Delta p \approx \Delta p_0 - \frac{2}{R_a}\left(H_e - \frac{2}{3}R_a\right)\tau_{wm} \tag{10.45}$$

In these equations

$$\tau_{wm} = \frac{\bar{\rho}gR_a}{2}(1 - \delta^*\phi)(1 - \Delta p^*) \tag{10.46}$$

The assumption of no slip at the wall should also be further investigated. It will be clear that when this assumption is not justified and there is slip this will increase the velocity of the dense phase in the annular zone and hence also in the core. Through glass windows in the wall of a 0.7 m diameter fluidized bed, wall slip velocities of the order of 1–2 m s^{-1} could be observed. It must be concluded that the phenomenon of wall slip of powders caused by shear stresses should be further investigated.

A final remark should be made concerning the assumption of constant porosity throughout the bed. At present there is no indication that this assumption is correct; in particular there is no reason to assume that the dense phase porosity in the central core is the same as that in the annulus. Hence, also the density of the dense phase may be different between core and annulus. These considerations, however, do not affect the principle of circulation and that of minimum energy dissipation.

REFERENCES

Abrahamson, A.R. & Geldart, D. (1980). Behaviour of gas-fluidized beds of fine powders. Part II. Voidage of the dense phase in bubbling beds. *Powder Techn.*, **26**, 47.

Bohle, W. & Van Swaay, W.P.M. (1978). The influence of gas adsorption on mass transfer and gas mixing in a fluidized bed. *Proc. Second Eng. Found Conf. on Fluidization*, Cambridge, ed. J.F. Davidson and D.L. Keairns. Cambridge University Press, London p. 167.

Carman, P.C. (1937). Fluid flow through granular beds. *Trans. Inst. Chem. Engrs*, **15**, 150.

Davidson, J.F. (1961). Symposium on fluidization-discussion. *Trans. Inst. Chem. Engrs*, **39**, 230.

Davidson, J.F. & Harrison, D. (1963). *Fluidized Particles*. Cambridge University Press.

Davies, L. & Richardson, J.F. (1966). Gas interchange between bubbles and the continuous phase in a fluidized bed. *Trans. Inst. Chem. Engrs*, **44**, T-293.

Harrison, D. & Leung, L.S. (1962). The rate of rise of bubbles in fluidized beds. *Trans. Inst. Chem. Engrs*, **40**, T-146.

Reuter, H. (1963). Druckverteilung um blasen in Gas-Feststoff-Fliessbett. *Chemie Ing. Techn.*, **35**, 98.

Reuter, H. (1965). Steiggeschwindigkeit von Blasen in Gas-Feststoff-Fliessbett. *Chemie Ing. Techn.*, **37**, 1062.

Rieke, R.D. & Pigford, R.L. (1971). Behaviour of gas bubbles in fluidized beds. *AIChE J.*, **17**, 1096.

Rietema, K. (1967). Application of mechanical stress theory to fluidization. *Proc. Int. Symp. on Fluidization*, Eindhoven, The Netherlands, p. 154.

Rietema, K. (1979). A critical review of the theory of potential flow of solids around rising bubbles in gas–solid fluidization in relation to the structure and rheology of the dense phase. *Chem. Eng. Sci.*, **34**, 571.

Rietema, K. & Ottengraf, S.P.P. (1970). Laminar liquid circulation and bubble street formation in a gas–liquid system. *Trans. Inst. Chem. Engrs*, **48**, T-54.

Rowe, P.N. & Matsuno, R. (1971). Single bubbles injected into a gas fluidized bed and observed by X-rays. *Chem. Eng. Sci.*, **26**, 923.

Rowe, P.N. & Partridge, B.A. (1965). An X-ray study of bubbles in fluidized beds. *Trans. Inst. Chem. Engrs*, **43**, T-157.

Rowe, P.N., Partridge, B.A. & Lyall, E. (1964). Cloud formation around rising bubbles in gas-fluidized beds. *Chem. Eng. Sci.*, **19**, 973.

Wijffels, J.B. & Rietema, K. (1972). Flow patterns and axial mixing in liquid–liquid spray columns. *Trans. Inst. Chem. Engrs*, **50**, 224, 233.

APPENDIX 10.A. ELABORATION OF THE CIRCULATION PROBLEM

The shear stress profile over the cross-section of the bed follows from the momentum equations. With the boundary condition that $\tau_d = 0$ for $r = 0$:

$$\tau_{dc} = -\frac{1}{2}\left\{\frac{dp}{dz} + (1-\varepsilon)(1-\delta^*)\rho_d g\right\}r \qquad \text{for } r < R_c$$

$$\tau_{da} = -\frac{1}{2}\left\{\frac{dp}{dz} + (1-\varepsilon)\rho_d g\right\}r + \delta^*(1-\varepsilon)\rho_d g\frac{R_c^2}{2r} \qquad \text{for } R_c < r < R_a$$

With $C_1 = (dp/dz) + (1-\varepsilon)\rho_d g$ and $C_2 = \delta^* \rho_d g (1-\varepsilon)$, it follows that

$$\tau_{da} = -\frac{1}{2}C_1 r + \frac{1}{2}C_2\frac{R_c^2}{r}$$

With $v_{da} = 0$ for $r = R_a$ the velocity profile in the annulus is found to be

$$v_{da} = \frac{C_1 r^2}{4\mu_d} - \frac{C_2 R_c^2}{2\mu_d}\ln\left(\frac{r}{R_a}\right) - \frac{C_1 R_a^2}{4\mu_d} \qquad (10.A.1)$$

After integration with respect to the radius the average velocity of the dense phase in the annulus is found to be

$$\bar{v}_{da} = C_1\frac{R_c^2 - R_a^2}{8\mu_d} + \frac{C_2 R_c^2}{2\mu_d}\left\{\frac{R_c^2}{R_a^2 - R_c^2}\ln\left(\frac{R_c}{R_a}\right) + \frac{1}{2}\right\} \qquad (10.A.2)$$

In the core the velocity is everywhere the same and equal to $v_{da}(r = R_c)$. Hence:

$$\bar{v}_{dc} = \frac{C_1(R_c^2 - R_a^2)}{4\mu_d} - \frac{C_2 R_c^2}{2\mu_d}\ln\left(\frac{R_c}{R_a}\right) \qquad (10.A.3)$$

We now introduce the dimensionless quantities β, ϕ, T_c and T_a:

$$\beta = \frac{C_1}{C_2} \qquad \phi = \frac{R_c^2}{R_a^2}$$

$$T_c = 4\mu_d v_{dc}/(C_1 R_a^2) = \phi\left\{1 - \frac{1}{\phi} - \frac{1}{\beta}\ln\phi\right\} \qquad (10.A.4)$$

$$T_a = 4\mu_d v_{da}/(C_1 R_a^2) = \phi\left\{\frac{1}{2}\left(1 - \frac{1}{\phi}\right) + \frac{1}{\beta}\left(\frac{\phi}{1-\phi}\ln\phi + 1\right)\right\} \qquad (10.A.5)$$

From the circulation balance of the dense phase it follows that

$$T_c(1 - \delta^*) + T_a\left(\frac{1}{\phi} - 1\right) = 0 \qquad (10.A.6)$$

Note that v_{da} is negative and, hence, also $T_a < 0$. Combination of eqns (10.A.4) – (10.A.6) gives:

$$\beta = \frac{(1 - \phi)/\phi + \delta^*\ln\phi}{\dfrac{1-\phi}{\phi}(1 - \delta^*) + \dfrac{1}{2}\left(\dfrac{1-\phi}{\phi}\right)^2} \qquad (10.A.7)$$

When we choose for a fixed case (determined by δ_o and v_{co}) a value for δ^* and a value for ϕ, then β is not a free choice any more but is determined by eqn (10.A.7). Besides, the gas balance (eqn (10.44)) has to be satisfied. With $v_s = 0$ this balance can be rewritten in the dimensionless form:

$$T_c = \frac{\lambda}{\beta\delta^*}\left\{\frac{1}{\phi\delta^*} - \frac{1}{\delta_o}\right\} \qquad (10.A.8)$$

in which $\lambda = 4\mu_d v_{co}/(\bar{\rho}gR_a^2)$ is the dimensionless fluidization velocity.

Variation of ϕ (with constant δ^*) will give us that specific value for which

$$(T_{c4} - T_{c8})/T_{c4} = 0$$

A value smaller than 10^{-4}, however, seems good enough. In this equation T_{c4} is the outcome of eqn (10.A.4) and T_{c8} the outcome of eqn (10.A.8).

With the relative bed expansion,

$$H_c/H_o = \frac{1}{1 - \phi\delta^*}$$

and the dimensionless pressure gradient,

$$-\frac{1}{\bar{\rho}g}\frac{dp}{dz} = (1 - \beta\delta^*)$$

we can find the dimensionless pressure drop,

$$\Delta p^* = \frac{(1 - \beta\delta^*)}{(1 - \phi\delta^*)}$$

With the specific value of ϕ and with the chosen value of δ^* the dimensionless pressure drop over the bed can be calculated. By varying δ^* (all at the same value of λ) the optimum value of δ^* for which Δp^* is a minimum can be found.

11

Handling of Powders

NOTATION

b_{ij}	Portion of the jth interval which falls by milling in the ith interval (—)
c_1	Dimensionless constant (—)
C	Composition of powder mix (—)
d_p	Particle size (m)
$\mathcal{D}$	Dispersion coefficient ($m^2\ s^{-1}$)
E	Elasticity modulus ($N\ m^{-2}$)
Fo	Fourier number (—)
g	Gravitational constant ($m\ s^{-2}$)
g^*	Resultant of gravitation and centrifugal acceleration ($m\ s^{-2}$)
l	Thickness of powder layer in the drum (m)
L	Part of the wall of the mill covered by powder, or length of the mixing drum (m)
n	Rotation speed (rpm) (s^{-1})
N	Number of balls inside the drum (—)
N_F	Fluidization number (—)
N_g	Particle–gas interaction number (—)
p	Pressure of gas ($N\ m^{-2}$)

229

q_{ci} Amount of powder of the ith interval milled by one cascading ball collision (—)

q_{fi} Amount of powder of the ith interval milled by one falling ball collision (—)

Q Amount of powder inside the drum (—)

R Radius of the ball mill (m)

R_k Radius of the milling balls (m)

s Spread of the slip velocity of balls

S_i Specific rate of breakage of powder of the ith interval (s^{-1})

t Mixing time (s)

t_d Time to discharge the upper hopper of the sand–glass (s)

U_c Velocity of a continuity wave (m s^{-1})

v_s Slip velocity of powder with gas (m s^{-1})

V_a Circumferential velocity of the mill (m s^{-1})

V_s Slip velocity of balls (m s^{-1})

w_i Weight fraction in the ith interval (—)

x Length coordinate (m)

α Dynamic angle of repose (—)

ε Porosity (—)

μ Gas viscosity (N s m^{-2})

ρ_d Density of the solids (kg m^{-3})

σ_{ci} $= q_{ci}/Qw_i$ (—)

σ_{fi} $= q_{fi}/Qw_i$ (—)

11.1 INTRODUCTION

In Chapter 1 it was mentioned that during handling of fine powders their mobility is increased by reshuffling so that extra gas is encapsulated and the powder can expand. The increased mobility of the powder strongly favours the desired action of the processing such as mixing and milling of the powder. In most cases the encapsulation of gas is caused by mechanical stirring of the powder in the process apparatus. As a consequence of the continuous action of gravity a pressure gradient arises which forces the entrapped gas to escape. Hence, the degree of expansion obtained depends strongly on the permeability of the powder and on the gas viscosity. In order to show the effect of the entrapped gas on the powder behaviour we did some experiments, described below.

11.1.1 Experiments in a Sand-glass

The sand-glass was made of perspex and consisted of two conical hoppers, one on top of the other, connected by an interchangeable nozzle between the two cones. The diameter of the nozzle could hence be varied (see Fig. 11.1). The two hoppers were also connected by a by-pass which could be closed. The gas atmosphere inside the sand-glass could be changed via two stopcocks. The sand-glass was filled with fresh cracking catalyst as a powder. When powder flows from the upper hopper to the lower hopper the pressure in the lower hopper must increase and hence gas tries to flow countercurrently with the powder through the nozzle or escapes through the by-pass when it is open.

With three different gases (H_2, N_2 and neon with viscosities 88×10^{-7}, 180×10^{-7} and 310×10^{-7} N s m^{-2} respectively), experiments were done with various nozzle diameters. The time t_d necessary to discharge all the powder (170 g) from the upper hopper was measured with the by-pass both open and closed. Fig. 11.2 shows that $(t_d)^{-0.5}$ is proportional to the nozzle diameter.

With the by-pass open the powder flows faster when the gas viscosity is lower, which seems reasonable. On the other hand when the by-pass is closed the situation is just the reverse and the powder flows faster (although always slower than with the by-pass open) when the gas viscosity is higher. At first sight this seems paradoxical. Since, however, in this case the gas must flow upwards through the nozzle, the powder in the upper hopper near the apex must be fluidized. As we have seen in Chapter 7 fluidization at higher gas viscosity results in higher porosity of the fluidized powder and this again in higher mobility. This experiment, therefore, clearly demonstrates that interaction of solids and gas is important and cannot be neglected.

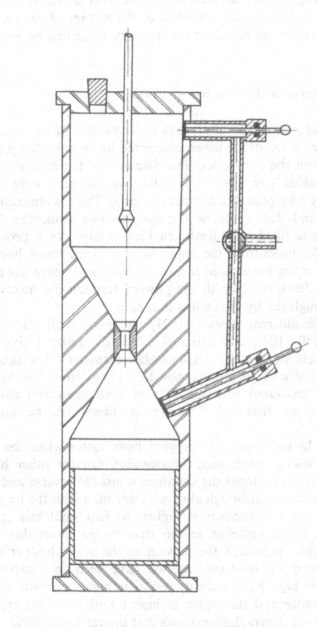

Fig. 11.1. Design of the sand-glass.

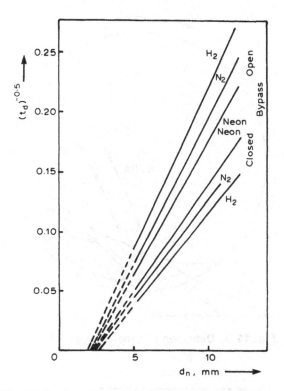

Fig. 11.2. Results of experiments in the sand-glass.

11.1.2 Experiments in a Two-dimensional Rotating Drum

In search of further evidence of gas–solid interaction in solids handling, we did experiments in a two-dimensional 'ball mill', again with fresh cracking catalyst. The mill had an internal diameter of 0·15 m and a thickness of 0·02 m. One side of this mill was of glass, providing the possibility of following the milling process visually by means of photographs at an exposure time of 0·001 s. The experiments discussed here were done without balls but at various gas pressures. In these experiments we studied the dependence of the dynamic angle of repose α_c of the powder as a function of the rotation speed of the drum. In Fig. 11.3 the dynamic angle of repose is defined. In Fig. 11.4 the results are presented. Again the effect of the gas viscosity is clear but the effect

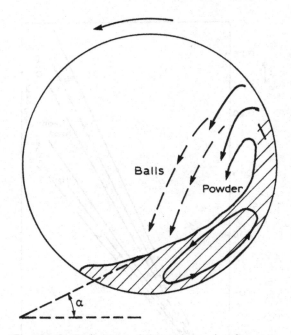

Fig. 11.3. Definition of dynamic angle of repose.

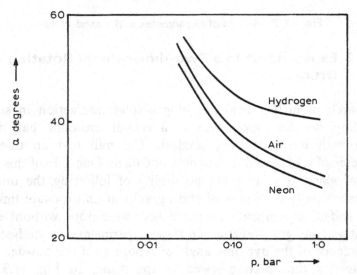

Fig. 11.4. Angle of repose versus gas viscosity and pressure.

of the gas pressure needs further explanation. Below 1 bar this influence can be understood when it is realized that the free path length of the molecules increases at decreasing pressure and below 0·1 bar becomes comparable with the size of the pores between the particles, which results in a decrease of the effective viscosity. Hence, below 0·1 bar the dependence of pressure can be explained as an effect of gas viscosity. At higher pressures, especially above 10 bar, the phenomenon of gas adsorption plays a role, which increases the interparticle forces. As was shown in Chapter 7 this favours higher bed expansion.

Although we have no data available, it was also clear from the experiments that when the angle of repose decreases at the same time the bulk volume of the powder increases due to the effect of the gas viscosity.

11.1.3 Criterion for Fluid Dynamic Gas–Solid Interaction

We shall now consider the reshuffled powder layer in the rotating drum more closely. The powder near the circulating wall of the drum will be continuously lifted up by the rotation; when it has reached a certain height, it rains down again in a powder shower towards the bottom (see Fig. 11.3). During this raining down the shower widens and hence entrains gas with it. This gas becomes engulfed into the powder layer on the bottom. Due to the action of gravity this gas tries to escape in the direction of the resultant g^* of the gravitational and centrifugal acceleration. This causes the powder to be fluidized when the engulfed gas cannot escape fast enough. A whole spectrum of continuity waves (see Chapter 7) arises and moves up. The velocity of these continuity waves is given by

$$U_c = -\frac{\mathrm{d}}{\mathrm{d}\varepsilon}\{\varepsilon(1-\varepsilon)v_s\} \qquad (11.1)$$

in which

$$v_s = \frac{\rho_d g^* d_p^2}{180\mu}\frac{\varepsilon^2}{(1-\varepsilon)}$$

Hence

$$U_c = -\frac{\rho_d g^* d_p^2}{60\mu} \varepsilon^2 \qquad (11.2)$$

Associated with each porosity value ε there is a specific continuity wave with a specific velocity which is the smaller the lower the porosity. As long as the continuity wave belonging to the packed bed porosity has not yet reached the surface of the powder layer, the more upward parts of the powder will have a porosity larger than this packed bed porosity. The time t_1 necessary for this continuity wave to reach the powder surface is l/U_c where l is the thickness of the powder layer. This time should be large compared with the time $t_2 = L/V_a$ where L is the part of the mill wall covered by the powder layer and V_a is the circumferential velocity of the rotating drum. With $l/L = \frac{1}{4}$ and $\varepsilon = 0.4$ this finally leads to the criterion

$$N_g = \frac{\rho_d g^* d_p^2}{\mu V_a} \ll 60\frac{l}{L}\frac{1}{\varepsilon^2} \simeq 100 \qquad (11.3)$$

It must be realized that this is a qualitative criterion which also depends on the type of apparatus in which the process is carried out and on the choice made of the critical velocity V_a. In practice, however, $N_g < 10$ will ensure that gas–solid interaction will play a significant role.

11.1.4 Conclusions

(1) By mechanical agitation of the powder environmental gas is entrapped into the powder bed. This causes the bulk volume of the powder and hence its porosity to be increased.

(2) The porosity increase is the higher the more difficult the entrapped gas can escape due to higher gas viscosity or to smaller particle size of the powder, as indicated by low values of the gas-influence number N_g.

(3) The porosity increase causes an increase of the powder mobility.

(4) As shown in Chapter 7 there is a maximum amount of gas that

a powder bed can contain when fluidized. Above this amount the powder structure becomes unstable and the excess gas escapes fast by bubbling. When by agitation of the powder, gas is entrapped, here also there is a maximum porosity above which the bed structure will not remain stable. Hence this maximum porosity cannot be exceeded. In fluidization this maximum porosity is determined by the fluidization number N_F according to

$$N_F = \frac{(\rho_d - \rho_c)^2 \rho_d d_d^4 g^2}{\mu^2 E} = \left\{ \frac{180(1-\varepsilon)}{\varepsilon^2(3-2\varepsilon)} \right\}^2$$

(see Chapter 7). By gas adsorption to the surface of the particles the interparticle forces increase, which causes an increase of the powder bed elasticity modulus E. It is believed that the same mechanism also operates when in powder handling the amount of engulfed gas becomes too high so that here also the excess gas will escape fast.

11.2 MILLING EXPERIMENTS

The milling apparatus was a laboratory-scale ball mill which could be operated with various gases and at pressures varying from 0·001 to 10 bar (see Fig. 11.5). Its internal diameter was 150 mm and its

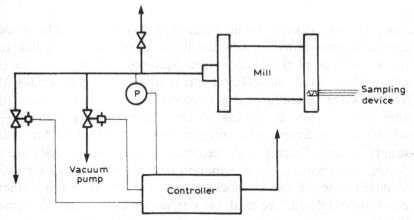

Fig. 11.5. Layout of laboratory-scale ball mill.

238 THE DYNAMICS OF FINE POWDERS

length 195 mm. The mill was filled with 240 steel balls with a diameter of 16 mm and was run at a speed of 80 rpm which is 73% of the critical speed. The bulk volume of the powder filling amounted to 15% of the mill volume. Samples of about 0·5 g were taken by means of a special sampling device consisting of a corkscrew closely fitting inside a hollow tube. This was inserted in the mill through a ball valve. By turning the corkscrew some powder was drawn into the tube (see Fig. 11.6). The whole

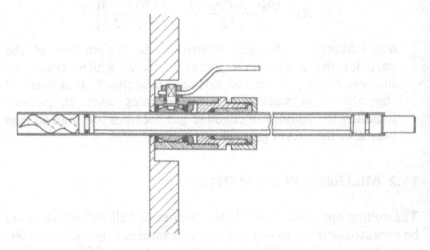

Fig. 11.6. Construction of sampling device.

assembly was airtight, so no leakage of gas was possible. Some contamination, however, was inevitable during sampling due to the air volume of the hollow tube. Although this volume equals only 0·05% of the mill volume, this is unacceptable, especially at low pressures. Therefore, the gas contents of the mill were renewed after each sample. By means of an airtight bearing, the mill—which was, of course, also airtight—was connected to the gas-supply system. During each experiment some 30 samples were drawn. Their particle size distribution was measured with a Coulter counter connected to a channel-analyser (see Fig. 11.7). For further experimental details the reader is referred to Cottaar and Rietema (1984).

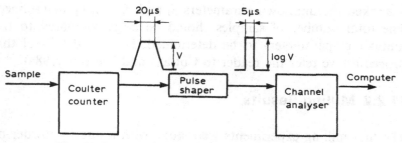

Fig. 11.7. Principle of processing of milling samples.

11.2.1 The Milling Model

To describe the milling process we applied the milling model as discussed in a review by Austin (1971) which is generally accepted by investigators in the field of milling. The weight size distribution of the powder is supposed to be split up into n intervals with a constant lower to upper size ratio. The nth interval contains all the particles below the upper size of this interval. The decrease with time of the weight fraction w_i of the interval i due to milling is assumed to be a first-order process:

$$\mathrm{d}w_i/\mathrm{d}t = -S_i w_i \qquad (11.4)$$

in which S_i is the specific rate of breakage of the ith interval. Furthermore, it is assumed that particles only get smaller and so do not agglomerate. The portion of the milled fraction of the jth interval which falls into the ith interval is denoted by b_{ij}, hence:

$$\mathrm{d}w_i/\mathrm{d}t = -S_i w_i + \sum_{j=1}^{i-1} b_{ij} S_j w_j \qquad (11.5)$$

while $S_n = 0$ because particles which fall into the nth interval remain inside this interval even when milled. Besides,

$$\sum_{i=j+1}^{n} b_{ij} = 1 \qquad (11.6)$$

because all milling products of size j have to fall into some higher interval.

When during the milling process enough samples of the powder are taken the unknown parameters S_i and b_{ij} can be determined. The total number of samples should be large compared to the number of parameters to be determined. For the details of this procedure we refer the reader to Cottaar and Rietema (1984).

11.2.2 Milling Results

The first milling experiments were done with a quartz powder of which the particle size distribution was split up into seven intervals with a higher to lower size ratio of 1·58 (see Table 11.1). Its cumulative weight fraction versus the interval number is shown in Fig. 11.8 together with the result after milling in air at a pressure

TABLE 11.1
Data on the Powders Investigated

Powder	ρ_d (kg m^{-3})	d_{p50} (μm)	N_g	Interval	Range (μm)	Weight fraction (%)
Quartz sand	2 600	69	69	1	89–140	13
				2	56–89	59
				3	35–65	19
				4	22–35	5
				5	14–22	2
				6	9–14	1
				7	9	1
Cracking catalyst	750	54	12	1	66–104	34
				2	42–66	48
				3	26–42	17
				4	17–26	1
				5	11–17	0
				6	7–11	0
				7	7	0
Hematite	4 900	40	44	1	56–89	15
				2	35–56	50
				3	22–35	25
				4	14–22	6
				5	9–14	2
				6	9	2

The N_g numbers were calculated for operation in a ball mill with a diameter of 150 mm and at a rotation of 80 rpm. The gas atmosphere was air.

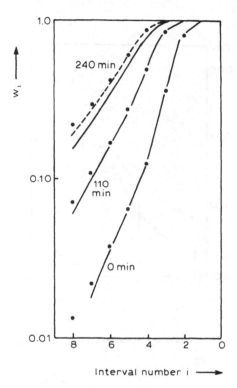

Fig. 11.8. Results of milling of quartz in air at 7 bar. The cumulative weight fraction as a function of the interval number at several moments, when milling in air at 7 bar.

of 7 bar during 110 and 240 min respectively. Milling experiments were further done at various gas pressures and in three different gas atmospheres (air, neon and hydrogen). The results were interpreted in terms of the milling parameters S_i and b_{ij} (see Figs 11.9 and 11.10). Again, strong effects of gas pressure and gas viscosity can be observed. It is believed that both these effects can be attributed to an effect of the powder mobility which increases with an increase of the porosity. In Fig. 11.11 the specific breakage parameters S_1 and S_2 obtained at a pressure of 1 bar and with three different gases are plotted versus the N_g number. This figure suggests a good correlation.

Milling experiments were also done with two other powders, viz.

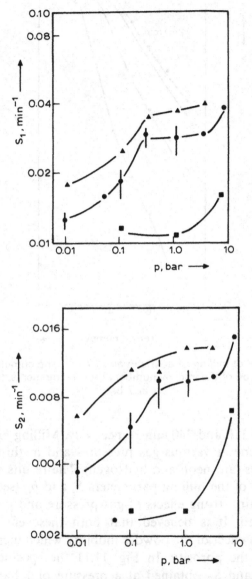

Fig. 11.9. The specific rates of breakage S_1 and S_2 as a function of pressure for three gases: ● air, ■ hydrogen and ▲ neon.

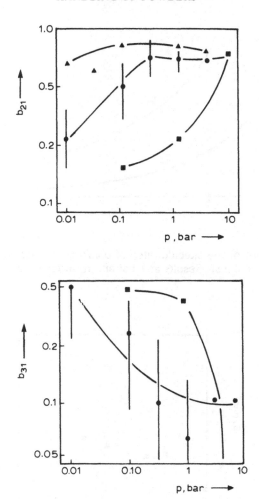

Fig. 11.10. The breakage parameters b_{21} and b_{31} as a function of the pressure for three gases: ● air, ■ hydrogen and ▲ neon.

fresh cracking catalyst and hematite (see Cottaar & Rietema, 1985). The relevant data for these powders are also given in Table 11.1. Since from earlier experiments it could be concluded that the effects of gas viscosity and gas pressure are almost similar, these latter experiments were done only with various pressures of air. Some of the results are given in Fig. 11.12 together with the results

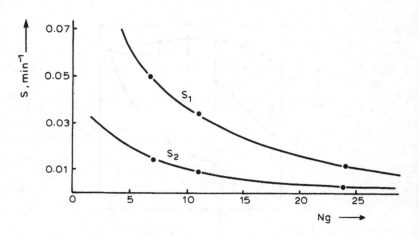

Fig. 11.11. Plot of the specific rates of breakage S_1 and S_2 versus the N_g number. Results at 1 bar and for three gases.

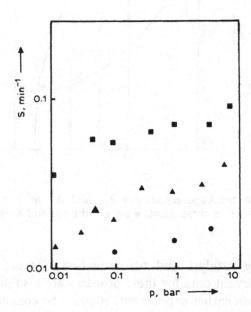

Fig. 11.12. The specific rates of breakage in air as a function of the pressure for three different powders: ▲ quartz sand, ■ cracking catalyst and ● hematite.

of milling quartz sand in air. These experiments provide further confirmation of the above conclusions.

For a further analysis of the milling process, Cottaar studied the behaviour of the balls inside the mill.

11.2.3 Visual Observation

The ball and powder behaviour during milling was observed visually in the two-dimensional rotating drum mentioned already in Section 11.1.2. The powder was fresh cracking catalyst and the balls were of a ceramic material. During milling photographs at an exposure time of 0.001 s were taken at three different gas atmospheres and at 0 and 15 min of milling, while also the rotation speed was varied ($n = 60$ and 80 rpm respectively) (see Fig. 11.13). Direct observation showed that in the ball behaviour roughly three types of ball movement could be observed, as follows.

(1) At the bottom of the mill rolling balls are observed. These balls are closely packed and move upwards. Generally, there is slip of these balls at the wall. The milling effect of these balls can be neglected.

(2) Above these rolling balls cascading balls can be seen. These balls hop down the slope of the powder layer and at each hop collide with another ball, so that at normal ball filling each ball can be assumed to experience R/R_k collisions with other balls (R_k is ball radius, R is mill radius).

(3) Finally there are the falling balls which are lifted up nearly to the top of the mill, to fall down in more or less parabolic trajectories over nearly the total diameter of the mill, thus giving a strong milling effect.

If V_a is the circumferential velocity of the mill and V_s is the slip velocity of the balls with respect to the wall, the number of balls lifted up per unit time is

$$n = c_1 N(V_a - V_s)/R_k \qquad (11.7)$$

in which c_1 is a dimensionless constant, which might depend on the ball filling but is of the order of unity; N is the number of balls in

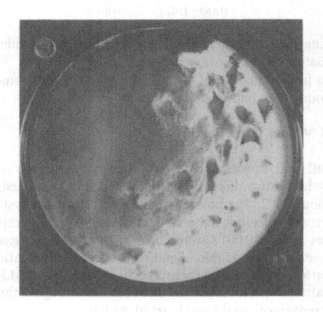

Air; t=0, n=60.

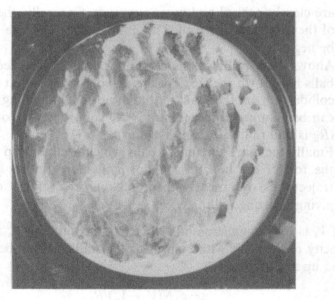

Air; t=0, n=80.

Fig. 11.13. Photographs of powder movement in the two-dimensional ball mill at various milling times (t) and rates of rotation (n) per minute for three gases: air, hydrogen and neon.

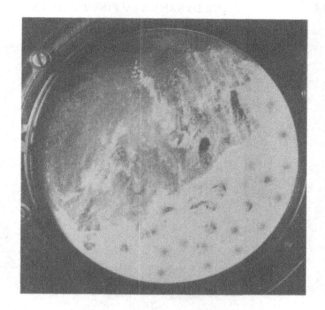

Air; $t=15$, $n=80$.

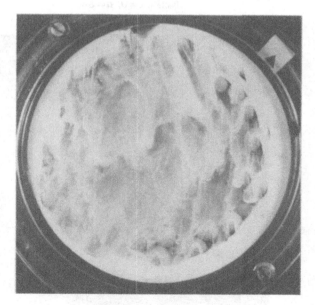

Hydrogen; $t=0$, $n=80$.

Fig. 11.13—*contd.*

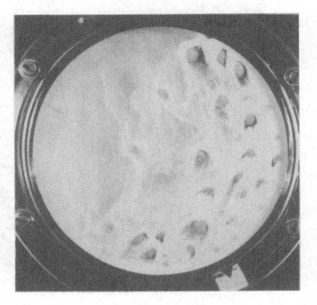

Neon; $t=0$, $n=80$.

Hydrogen; $t=15$, $n=80$.

Fig. 11.13—*contd.*

the drum; and V_s is an average value since not all the balls have the same slip velocity. Balls with too high a slip velocity do not acquire enough kinetic energy to become falling balls and hence will come down as cascading balls. Only balls with a slip velocity smaller than some critical velocity V_{sc} will ultimately convert into falling balls. Assuming a Gaussian distribution of the ball slip velocity, the number of falling balls will be given by

$$n_f = c_1 \frac{N}{R_k} \frac{V_a - \bar{V}_s}{s\sqrt{2\pi}} \times \int_0^{V_{sc}} \exp\left[-\frac{(V_s - \bar{V}_s)^2}{2s^2}\right] dV_s \quad (11.8)$$

while the number of cascading ball collisions follows from

$$n_c = \frac{c_1 N R}{R_k^2} \frac{V_a - \bar{V}_s}{s\sqrt{2\pi}} \times \int_{v_{sc}}^{\infty} \exp\left[-\frac{(V_s - \bar{V}_s)^2}{2s^2}\right] dV_s \quad (11.9)$$

The ball behaviour was studied at various ball loads. The rate of falling ball collisions n_f could be determined by means of an analysis of the noise produced by the mill. The rate of cascading ball collisions followed from an analysis of the power consumption by the mill at various ball loads. For further details reference should be made to Cottaar and Rietema (1986). Table 11.2 gives the results of these experiments. They indicate that at increasing gas viscosity (decreasing N_g number) the rate of falling ball collisions decreases. This must be due to an increase of the ball slip velocity V_s when the powder mobility increases. At the same time the rate of cascading ball collisions increases but probably reaches a maximum.

TABLE 11.2
Falling-Ball and Cascading-Ball Rates at Varying Ball Loads and Two Gas Pressures

N	0·01 bar, $N_g = 22$		1·0 bar, $N_g = 11$	
	n_f	n_c	n_f	n_c
60	0	780	0	490
120	0	1 480	0	980
180	50	1 860	0	1 870
240	165	1 620	10	2 670
360	270	300	90	1 200

Suppose that q_{fi} is the amount of powder from the ith interval milled by one falling ball collision and q_{ci} is the amount of powder from the ith interval milled by one cascading ball collision. If Q is the total amount of powder present in the mill, we introduce

$$\sigma_{fi} = q_{fi}/Qw_i \text{ and } \sigma_{ci} = q_{ci}/Qw_i$$

It follows that

$$S_i = n_f \sigma_{fi} + n_c \sigma_{ci} \tag{11.10}$$

Of course, σ_{fi} and σ_{ci} strongly depend on the powder characteristics and on the geometry of the ball mill, but keeping these parameters constant and varying only the ball load, no variation of σ_{fi} and σ_{ci} should be expected. Hence, when S_i, n_f and n_c are known, σ_{fi} and σ_{ci} can be determined. The results are given in Table 11.3. As could be expected, σ_{fi} and σ_{ci} increase with

TABLE 11.3
Milling Parameters at Three Intervals and Two Gas Pressures, $N=240$

	Interval	$\sigma_{fi} \times 10^6$	$\sigma_{ci} \times 10^8$	$n_c \sigma_{ci}/S_i$
1 bar, $N_g=11$	1	13 +2	33 +3	0·872
	2	5·4+0·8	8·9+0·9	0·815
	3	1·9+0·4	3·0+0·4	0·809
0·01 bar, $N_g=22$	1	2·5+0·2	5·4+0·1	0·175
	2	0·7+0·1	1·0+0·1	0·123
	3	0·22+0·05	0·34+0·08	0·132

decreasing N_g number. This is due to the fact that before they collide with other balls or with the wall of the mill, balls lose kinetic energy by frictional dissipation with the surrounding powder. The higher the powder mobility the smaller is this energy loss and the more energy is saved for the final collision. It seems reasonable that cascading balls experience more hindrance from the surrounding powder than do falling balls. Also the collision energy of cascading balls is much lower than that of falling balls. This explains why $\sigma_{fi} \gg \sigma_{ci}$.

11.2.4 Scaling-up

When scaling-up the milling operation the radius of the ball mill R must increase and, since ball mills are generally operated at a rotation speed some 20% below the critical speed, this means that V_a must increase also and proportionally with the square root of R. Hence the N_g number will decrease. Assuming that V_s increases at the same ratio as V_a it follows from eqn (11.8) that n_f/N will increase with $\sqrt{R}$. Due to the decrease of the N_g number, however, the powder mobility will increase. This causes V_s to increase strongly and the overall effect will be a decrease of the number of falling balls. At the same time the number of cascading balls increases and their contribution to the overall milling effect increases. For the first interval:

$$\frac{\text{milling effect of cascading balls}}{\text{overall milling effect}} = n_c \sigma_{c1}/S_1$$

$$= 0.175 \text{ at } N_g = 22$$
$$= 0.872 \text{ at } N_g = 11$$

Hence, when the conditions during milling are (or become) such that $N_g < 20$ the contribution of falling balls to the overall milling effect reduces; at further scaling-up it finally becomes negligible and the milling result will be determined mainly by the cascading balls (see Fig. 11.14).

From the foregoing it follows that it makes sense to apply rough or ribbed linings to the inside wall of the mill. This will reduce the slip velocity of the balls so that the number of falling balls will increase. Besides, these balls will be lifted up higher so that their potential energy will increase as well as their final collision energy, resulting in an increase of q_{fi}.

Scaling-up of ball mills has also been reported in the literature, e.g. by Austin (1973) and Gupta *et al.* (1985). All their results were obtained, however, with powder–mill systems for which $N_g > 100$, so that from the theory developed here it can be concluded that the breakage rates as found by Austin and Gupta were mainly due to the effect of falling balls and hence cannot be compared with this investigation.

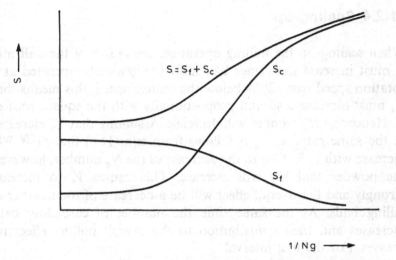

Fig. 11.14. Contribution of falling balls S_f and cascading balls S_c as functions of the reciprocal of the N_g number.

All our results were obtained while the ball mill was operated batchwise. In these experiments the N_g number decreased continuously but was less than 20 from the start.

In practice ball-mill operation will often be continuous, the rough powder being fed in at one end of the rotating drum while the product is taken off at the other end of the drum. Hence, there arises an axial gradient of the size of the powder particles and so also of the powder mobility. At the discharge end of the drum the bulk density of the powder will probably be lower than at the feed end. This will have the effect of an increased axial mixing: the rough powder with high bulk density will slip below the milled powder with low bulk density. Thus the milled powder might become polluted with traces of rough unmilled powder.

11.3 MIXING AND SEGREGATION

Mixing of two powders A and B is generally carried out in an apparatus by stirring the powder mechanically. This means that these powders are dispersed in each other such that at the end of the process in all arbitrarily small but equal volumes of the mixture an equal amount Δa of powder A and an equal amount Δb of powder B will be found. In practice this aim can never be reached because of the corpuscular character of powders which prevents the above volumes from being taken as arbitrarily small.

Segregation of a mixture of two powders can occur when the mobility of the mixture is high enough to overcome possible cohesion and when the powders are different in particle size and/or particle density. It may then happen that the powder with the lower density (or the smaller particle size) will be driven out of the mixture in the direction against gravity (or centrifugal force) and will collect at the top of the powder layer, while the other powder collects at the bottom.

The segregation process is further complicated by the outflow of the gas which is continuously engulfed into the mixture as the consequence of the mechanical stirring. As this gas tries to escape from the powder it exerts an upward force (the slip-force) on the particles of the powder; and, of course, particles moving in different directions also exert slip-forces on each other.

The driving force of the segregation is the buoyancy of the particles in the mixture minus their weight, which force makes equilibrium with the above-mentioned slip-forces. The buoyancy (see Chapter 5, Section 5.4) is a function of the local composition of the mixture and of the local porosity. The mixing parameter will also be a function of the composition and of the porosity. Since segregation and mixing will always be in competition with each other, the whole process will be extremely complicated.

11.3.1 Experiments

In order to determine the true mixing rate and the effect of the interstitial gas on the mixing, we decided to eliminate the segregation by studying the mixing of a powder with itself (see Cottaar & Rietema, 1986). To that end we chose fresh cracking catalyst (FCC) as a powder, part of which was marked with a suitable tracer which does not affect the flow properties of the powder. As FCC is a white powder an obvious tracing method was painting with black ink. To ensure that the relevant properties (fluidization behaviour, particle size distribution, etc.) of painted and unpainted powder were practically the same, the unpainted powder was treated in the same way as the painted powder except that no ink was added (i.e. it was also wetted in alcohol, dried and sieved again). The experiments were carried out in a horizontal rotating drum which was only partly filled with the powder: at the start one half with white powder, the other half with coloured powder (see Fig. 11.15). In order to reduce the slip velocity of the powder at

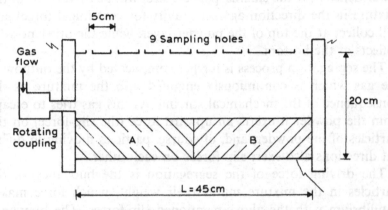

Fig. 11.15. Mixing black and white fresh cracking catalyst in a horizontal rotating drum (no balls).

the cylindrical wall of the drum, this wall was covered with a thin layer of felt. The gas handling system used was much the same as that used in the milling experiments discussed in Section 11.2. The rotating drum was fitted with nine sampling holes. After a certain mixing time t the rotation was stopped and at each sampling hole an average composition was obtained by measuring at a number of random radial positions by means of the optical fibre probe which is schematically indicated in Fig. 11.16.

To interpret the results it was assumed that the mixing process could be described by means of a diffusion equation:

$$\frac{\partial C}{\partial t} = \frac{\partial}{\partial x}\left\{ \mathscr{D} \frac{\partial C}{\partial x} \right\}$$

(11.11)

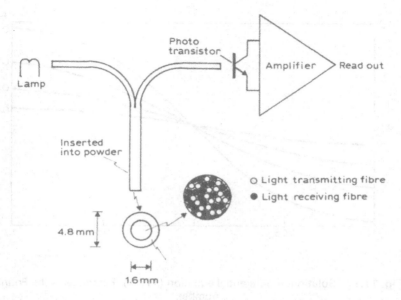

Fig. 11.16. Optical fibre probe to measure the composition of the catalyst powder.

in which x is the axial coordinate and C the composition defined as $C = \Delta a/(\Delta a + \Delta b)$. $\mathscr{D}$ is the dispersion coefficient. The initial conditions are:

$$\text{at } t = 0 \quad C = 1 \quad \text{for} \quad 0 < x < L/2$$
$$C = 0.5 \quad \text{for} \quad x = 0$$
$$C = 0 \quad \text{for} \quad -L/2 < x < 0$$

The boundary conditions are:

$$\mathrm{d}C/\mathrm{d}x = 0 \quad \text{for} \quad x = \pm L/2$$

The solution of this equation is given in Fig. 11.17 where the Fourier number $\mathrm{Fo} = \mathscr{D}t/L^2$ is the parameter. In Fig. 11.18 two examples of measured composition distributions are given and compared with the theoretical distributions for two values of F_0, 0·066 and 0·175 respectively. By means of curve-fitting with Fig. 11.17, the Fourier number pertaining could be determined. The dispersion coefficient then follows by multiplication by L^2/t. Figure 11.19 gives the dispersion coefficient thus determined as a

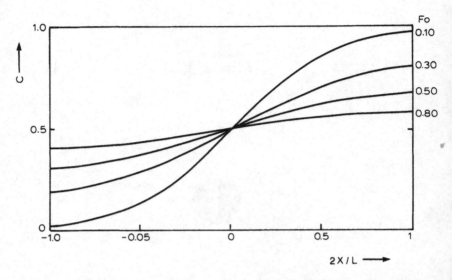

Fig. 11.17. Solution of differential equation (11.11). Parameter is the Fourier number.

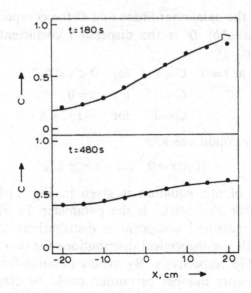

Fig. 11.18. Measured composition of powder when mixing in argon at 1·8 bar at a rotational frequency of 60 rpm after 180 and 480 s respectively.

function of the gas viscosity at 1·8 bar and 60 rpm. Finally, Fig. 11.20 gives the dispersion coefficient with air and as a function of the gas pressure, also at 60 rpm. The results conform well with the theory developed in this chapter.

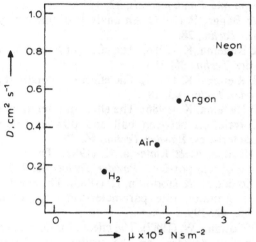

Fig. 11.19. Dispersion coefficient as a function of gas viscosity. Measured in the catalyst powder with various gases at a pressure of 1·8 bar and a rotation frequency of 60 rpm.

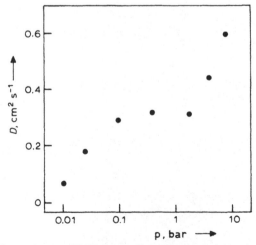

Fig. 11.20. Dispersion coefficient as a function of gas pressure when mixing in air at 60 rpm.

REFERENCES

Austin, L.G. (1971). A review. Introduction to the mathematical description of grinding as a rate process. *Powder Techn.*, **5**, 1.

Austin, L.G. (1973). Understanding ball mill sizing. *Ind. Eng. Chem. Process Design and Fundamentals*, **12**, 121.

Austin, L.G. & Bagga, P. (1981). An analysis of fine dry grinding in ball mills. *Powder Techn.*, **28**, 83.

Cottaar, W. & Rietema, K. (1984). The effect of interstitial gas on milling. Part I. *Powder Techn.*, **38**, 183.

Cottaar, W. & Rietema, K. (1985). The effect of interstitial gas on milling. Part II. *Powder Techn.*, **43**, 189.

Cottaar, W. & Rietema, K. (1986). The effect of interstitial gas on milling. Part III. Correlation between ball and powder behaviour and the milling characteristics. *Powder Techn.*, **46**, 89.

Cottaar, W., Heynen, A. & Rietema, K. (1987). The effect of interstitial gas on mixing of fine powders. *Powder Techn.*, **46**, 219.

Gupta, V.K., Zouiet, H. & Hodouin, D. (1985). The effect of ball and mill diameters on grinding rate parameters in dry grinding operation. *Powder Techn.*, **42**, 199.

Rietema, K. & Cottaar, W. (1987). The effect of interstitial and circumambient gas in fine powders on the scaling up of powder handling apparatus as illustrated by ball mill operation. *Powder Techn.*, **50**, 147.

Index

Printed in the United States
By Bookmasters